T0251158

PROJECT AND POLICY EVALUATION IN TRANSPORT

Project and Policy Evaluation in Transport

Edited by
LIANA GIORGI
ALAN PEARMAN
with
ANNURADHA TANDON
DIMITRIOS TSAMBOULAS
CHRISTIAN REYNAUD

Routledge
Taylor & Francis Group

LONDON AND NEW YORK

First published 2002 by Ashgate Publishing

Reissued 2018 by Routledge
2 Park Square, Milton Park, Abingdon, Oxon OX14 4RN
711 Third Avenue, New York, NY 10017, USA

Routledge is an imprint of the Taylor & Francis Group, an informa business

A Library of Congress record exists under LC control number: 2001098006

ISBN 13: 978-1-138-72892-9 (hbk)
ISBN 13: 978-1-315-19023-5 (ebk)

Contents

Notes on Contributors   *vi*
Foreword   *ix*

1   Introduction: The Theory and Practice of Evaluation
    Liana Giorgi and Annuradha Tandon   1

2   The Policy-Making Process in the European Union
    Francis McGowan   14

3   The EU Enlargement and its Impact on European Policies
    Gerhard Rambow   53

4   Enlarging EU Environmental Policy: The Challenges of
    Flexibility and Integration
    Ingmar von Homeyer   99

5   Analytical Frameworks for Policy and Project Evaluation:
    Contextualizing Welfare Economics, Public Choice and
    Management Approaches
    Wayne Parsons   144

6   Evaluation of Projects and Programmes: Principles and
    Examples
    Frank Haight   181

7   Transport Evaluation Methods: From Cost-Benefit Analysis
    to Multicriteria Analysis and the Decision Framework
    Michel Beuthe   209

8   Criteria for Evaluation Towards Sustainability
    Klaus Rennings and Sigurd Weinreich   242

Notes on Contributors

Michel Beuthe received a D. in Law (UCL, Belgium) and a Ph.D. in Economics (Northwestern, USA). He is a Professor at the Facultés Universitaires Catholiques de Mons (FUCAM), Belgium and the Research Director of Groupe Transport and Mobilité (GTM). His research focuses on Cost-Benefit and Multicriteria Analyses of Public Investments, particularly of transport infrastructures, with an emphasis on the treatment of uncertainty, and secondly on Network Analysis of Multi-Modal and Intermodal Networks of Transportation, particularly for freight. The results of these researches have been published in several referred journals and books.

Liana Giorgi is Vice-Director of the ICCR and responsible for a number of international projects run by the institute in the departments of *Social Policy Analysis* and *Transport Policy Analysis and Evaluation*. She is co-editor of the ICCR book series *Contemporary Trends in European Social Sciences*; and of *Innovation – The European Journal of Social Sciences*. She previously worked at the Institute of Women's Studies at the University of Amsterdam and received her degrees in social and political sciences at the University of Cambridge; and a degree in cognitive science from MIT (USA). She was national co-ordinator of the European Community Household Panel (ECHP) in Austria between 1994-1998; scientific co-ordinator of the project CODE-TEN 'Strategic Assessment of Corridor Developments' (1998-1999) and research manager of the Thematic Network 'Policy and Project Evaluation' (1999-2000). She is the author of *The post-socialist media: What power the West?* (Avebury, 1995), and co-editor of *European Transformations: Five decisive years at the turn of the century* (Avebury, 1995) in addition to numerous journal articles on the above topics.

Frank Haight has studied traffic safety for thirty-five years. His publications deal with many aspects of the subject, including truck regulation, speed and speed limits, alcohol and drinking driving, the relationship of safety with mobility, accident proneness, suicide, induced exposure, alertness, data systems, accident epidemiology, developing countries' accidents, pedestrian safety, fatality and injury rates, project evaluation, risk analysis, international co-operation, intervention analysis, cost/benefit analysis, and safety policy. Since 1969, he has been Editor-in-Chief of the international journal *Accident Analysis and Prevention* published by Elsevier Science of Amsterdam.

Ingmar von Homeyer is a research fellow at Ecologic, Institute for International and European Environmental Policy, Berlin. His work focuses on European environmental policy, the environmental policy implications of EU enlargement, and European and international regulation of biotechnology. His most recent publications include 'Flexibility or Renationalisation: Effects of Enlargement on EC Environmental Policy', in M.G. Cowles and M. Smith (eds), *State of the European Union: Risks, Reform, Renewal and Revival*, Vol. 5, Oxford: Oxford University Press, 2000 (with S. Bär and A. Carius); 'Overcoming Deadlock? Enhanced Co-operation and European Environmental Policy after Nice' in Somsen, H. et al. (eds), *The Yearbook of European Environmental Law*, Vol. 2, Oxford: Oxford University Press, forthcoming (with S. Bär and A. Klasing); 'A Nice Environment for Enlargement', *European Environmental Law Review*, forthcoming (with S. Bär and A. Klasing).

Francis McGowan is a lecturer in Politics at the School of European Studies, University of Sussex. His research interests include energy and transport policy and the political economy of regulation in the UK and the EU. His publications include *The Struggle for Power in Europe* (RIIA, 1993) (as editor and contributor) *European Energy Policy in a Changing Environment* (Springer, 1996) and (as co-editor and contributor) *Social Democracy – Global and National Perspectives* (Palgrave, 2001). He has acted as a consultant to the European Commission as well as a number of firms in the energy sector.

Wayne Parsons is Professor of Public Policy and Head of the Department of Politics at Queen Mary, University of London. He is editor of the series '*New Horizons in Public Policy*' for Edward Elgar Publishers, and is on the editorial board of the '*Journal of European Public Policy*' and '*Contemporary Wales*'. Professor Parsons was the founding director of QMW's Public Policy Research Unit and its post-graduate programme in public policy.

Gerhard Rambow was Director General for European Affairs in the German Federal Ministry of Economics in Bonn until mid-1996. He has been active as an independent consultant in countries of Central and Eastern Europe since then, advising them on their course to joining the EU. He served in the Ministry for Economics for 35 years and focused on questions regarding European integration, beginning in 1964. Harmonisation of law, completion of the internal market, structural funds, control of state aid, negotiations of the Europe Agreements, and in general, the coordination of European Policy were among his tasks. He also taught a course in Law of Competition at the University of Wuppertal until mid-1997.

Klaus Rennings Since 1994, Klaus Rennings has been a senior researcher at the ZEW, Department of Environmental and Resource Economics, Environmental Management. He studied at the University of Münster. From 1990 to 1994, he worked at the University of Münster as a researcher at the Institute for Transport Economics. From 1992 to 1994 he was an assistant to Prof. Dr. Ewers at the German Council of Environmental Advisors, an advisory council for the Federal Government comprised of environmental scientists from several disciplines. His main fields of research include the valuation of environmental risks and the operationalisation of the concept of sustainable development by monetary and physical indicators. His dissertation was published in 1994 with the title *Sustainability Indicators*. Other publications cover the fields of external costs and a regulatory framework for a policy of sustainability, including the economic assessment of environmental policy instruments. His current field of research includes determinants and impacts of environmental innovations. He is especially interested in the role of regulation stimulating eco-innovation and in the employment effects of cleaner production.

Annuradha Tandon is a Scientific Researcher with the ICCR. She is an Economist from the Delhi University and a Post Graduate in Business Administration. Her research interests focus on Transport Policy Analysis and Evaluation and Urban Transport. She has been actively involved in a number of projects under the Fourth and Fifth Framework programme of the DG Transport.

Dipl.Vw. Sigurd Weinreich, Environmental Economist, has studied economics at the University of Heidelberg, Germany. Since 1995, he has been a research fellow at ZEW, Department of Environmental and Resource Economics, Environmental Management. His main field of research is transport and environment. Mr. Weinreich collaborated on the EU-project QUITS (Quality Indicators of Transport Systems), evaluating the quality of transport systems on three European long distance routes. Currently he is working on the follow-up project RECORDIT (Real Cost Reduction of Door-to-door Intermodal Transport) leading the methodological part of internal and external cost calculation. In 1998 and 1999, he was a member of the EU Concerted Action CAPRI (Concerted Action for Transport Pricing Research Integration in Europe). His doctoral thesis will be in the field of sustainable mobility with a focus on external costs of passenger transport.

Foreword

This book presents the results of the seminar 'Policy and Project Evaluation: Context, Theory and Methods' organised by the Interdisciplinary Centre for Comparative Research in the Social Sciences (ICCR) in Brussels in May 2000. The seminar was organised with the support of the European Commission (DG Energy and Transport) in the framework of the thematic network TRANS-TALK on Policy and Project Evaluation Methodologies in Transport of the Fifth Framework Programme (Key Action 'Sustainable Mobility and Intermodality').

The editors would like to thank the European Commission for their support of the TRANS-TALK thematic network. We are particularly grateful to Catharina Sikow-Magny of DG-TREN for her considerate supervision of our work throughout its course.

We are also all indebted to Niki Rodousakis and Johannes Blaas of the ICCR for their assistance with copy editing and the preparation of the final manuscript for the publishers.

The Editors

1 Introduction: The Theory and Practice of Evaluation

LIANA GIORGI AND ANNURADHA TANDON

Evaluation is not a new field: it grew in the 1950s in parallel with the implementation of large-scale programmes for urban and regional development following the end of World War II and gradually gave rise to a new specialisation within the social sciences, namely policy and public administration specialists or policy analysts (cf. Rossi and Freeman, 1993). There is by now a wide literature on the principles of evaluation, its objectives, the methods, the problems and barriers and its use in policy formulation and implementation (for a review of the relevant literature see OECD – Freeman, Rossi and Wright, 1980; Rossi and Freeman, 1993; Parsons, 1995; and Chelimsky and Shadish, eds, 1997). Evaluation is in several countries well ingrained in the policy decision-making process: forerunners in this connection are the U.S. and the U.K. where the General Accounting Office and the Treasury respectively regularly produce guidelines for evaluation.

Evaluation is also not new in the transport sector. However it is not as entrenched at the level of policy as in the fields of health, education, housing or work. There are several reasons for this. Indicative is that the term 'policy analysis' as a research field in its own right is indeed new in transport. Instead the terms 'planning' and 'assessment' are dominant, with infrastructure investment – and hence also project appraisals – delineating the main reference framework. This has also meant that the professional group that has emerged to provide evaluations comprises mainly economists and engineers with little input from political scientists or institutional economists who have dominated the policy evaluation field in other areas. Furthermore, even though several transport economists argue in favour of a specialised approach within economics for transport 'in many respects transport economics is simply the application of microeconomic principles and methods to an economic activity consisting of the movement of freight and passengers' (Rus and Nash, 1997, p.1).

1

In fields like health or education, policy programmes or local projects explicitly and directly target the welfare of citizens or seek to address specific social needs. Here evaluation is called upon to assess these needs and/or the outcomes of the intervention. Transport is most often seen as a derived demand, a service not required in itself, but simply to facilitate the meeting of other human needs. It is hence not surprising that in the field of transport planning the objectives more often than not relate to economic efficiency and/or growth, and evaluation is applied for 'checking plans for public expenditures' (Rus and Nash, 1997); for estimating time savings; for investigating mainly at the macro-economic level the relation between infrastructure investment and urban or regional development (cf. Banister and Lichfield, 1995); or for assessing social and environmental impacts, albeit as externalities (cf. Hoon Oum *et al.*, 1997).

The process of European integration has however changed the evaluation landscape in transport – the Transport RTD programme under the Fourth Framework Programme provides ample evidence to this (cf. Banister and Berechman, 1993; Giorgi and Pohoryles, 1998). The drive towards harmonisation has brought policy evaluation as distinct from project appraisal onto the agenda and has increased demands for 'strategic assessments' for checking the consistency of 'policies, plans and programmes (P/P/P)' or for approaching 'the design of projects in a generic sense' (EC, 1994; EC, 1997a; Banister and Lichfield, 1995). A recent European Initiative – the Sound and Efficient Management 2000 Programme – has sought to elaborate a framework for carrying out evaluations of Community programmes (cf. EC, 1997b; EC, 1999) and has increased awareness of the significance of the evaluation function in policy formulation and deliberation.

These new demands in the field of transport evaluation arise by way of the debate on 'sustainability' both with respect to the environment and with respect to distributional considerations or accessibility. This is not surprising. These are issues that question the utilitarian principles that underlie the classical transport planning approach with its strong emphasis on economic efficiency. From the methodological viewpoint they also reveal the complexity of impact assessment where there is a multitude of types of impacts and impact groups and where cost-benefit or cost-effectiveness analyses are required.

This book represents the first attempt in transport literature to address three key questions for evaluation raised by the afore-mentioned challenges and new demands.

- What is the new *European* context of evaluation? Three contributions deal with three important areas: the institutional context of transport policy-making and implementation is the subject of the contribution of *Francis McGowan*; the developments in environmental policy that of *Ingmar von Homeyer*; and the challenges of enlargement the topic of *Gerhard Rambow*.
- How relevant is evaluation theory for addressing contemporary transport policy challenges? Two contributions remind the reader of the importance of theory and related to this of two obvious (yet frequently forgotten) scientific facts: first, that methods are not neutral but rely on analytical frameworks which, in turn, build on specific assumptions about epistemology, methodology, complexity, institutions and welfare – this is the subject of the contribution of *Wayne Parsons*; second, that the use of quantification is not equivalent to blind positivism in the sense of rendering social theory and conceptual or methodological choices obsolete – the topic of the contribution of *Frank Haight*.
- How useful are conventional assessment methods like cost-benefit analysis or multi-criteria analysis for answering modern complex evaluation questions in transport? Still useful but not unconditionally so and not without further elaboration, argue both *Michel Beuthe* and *Klaus Rennings* (with *Sigurd Weinreich*) in their respective contributions. What is today needed is a decision framework comprising several parameters. Cost-benefit analysis continues to provide important information in this respect but is alone not enough. This is true generally but particularly with regard to the inclusion of new types of impacts like environmental effects.

Clearly one book can alone not cover all the relevant issues and not equally in depth. There are several other issues central to transport evaluation at European level which remain unexplored. A second book currently under preparation by some of the editors of this book aims to advance the discussion further. The two books should be read in sequence but represent also independent achievements. Hopefully what this first book will have achieved is to locate the discussion in an appropriate framework, namely in a framework which is specific to transport yet at the same time generic enough to allow a comprehensive reflection on the issues at hand thus avoiding the pitfalls of over-specialisation.

In the remainder of this introduction we provide a short introduction to key terms and issues in evaluation and a summary of the main arguments made by the contributors to this book.

Why Evaluation?

Evaluation is 'a process which seeks to determine as systematically and objectively as possible the relevance, efficiency and effect of an activity in terms of its objectives' (Rossi and Freeman, 1993, p.3). In other words it represents the assessment of the outputs, outcomes and processes of an activity (US General Accounting Office, 1991).

When tied to the original goals and objectives of public policy, evaluation represents a process of verification of the extent to which these goals were met as well as a procedure for improving the accountability of public institutions. Following from the above, it carries a normative element and has a practical orientation as it is expected to produce results that can be applied to improve policy interventions (EC, 1999).

Evaluation can have one or several objectives. It can be employed for judging whether an intervention is legitimate or not; for examining whether an activity conforms to statutory and regulatory requirements, programme designs and professional standards; to provide feedback as part of a monitoring exercise; or for assessing the outcomes of a policy intervention and, in this connection, to provide information on the use and allocation of public resources or the efficiency of a programme.

Scope of Evaluation

No evaluation exercise can meet all of the above objectives at the same time. This is why it is useful to determine the scope of an evaluation exercise early in the evaluation process.

The scope of the evaluation is determined by the decision-making process. There are two main aspects to this: the first is the stage of the decision process at which evaluation is employed; the second is the level of analysis.

With reference to the stage of the decision process, a common reference point is that of the classical decision-making model which distinguishes between four main phases, namely the agenda-setting or problem definition stage, the policy design phase, the policy legitimation phase and the policy implementation phase. Palumbo (1987) reformulates the objectives of evaluation according to this policy cycle as follows:

- In the agenda setting and problem-definition phase, the objective of evaluation is to define the size and distribution of the problem, to forecast or determine the needs, and to identify the target groups or areas.
- In the policy design phase, decision analysis techniques involve the identification of alternative means of achieving programme ends with the purpose of selecting the most cost-effective alternative.
- In the policy legitimation phase, evaluation must assess the acceptance of a policy or programme by the public and stakeholders.
- Finally in the policy implementation phase, evaluation checks whether the policy is implemented properly, i.e. according to standard procedures and in line with the original objectives.

In transport evaluation literature (Leleur, 1995; Turro, 1999; Sugden and Williams, 1978; Layard and Glaister, 1996; Atkinson and Cope, 1994; Pearce and Hett, 1999) one refers to *ex ante evaluation*, also known as appraisal, to describe evaluation that is carried out in the problem definition or policy design phases; *intermediate evaluation* to describe evaluation undertaken during the implementation stage for the purpose of monitoring; and *ex post* evaluation to describe evaluation that is carried out once the implementation phase is completed and for the purpose of assessing the project or programme's impacts.

Insofar as the level of analysis is concerned, it is widespread practice to distinguish between policies, programmes and projects, where policy refers to a set of programmes or measures that have the same specific objectives, schedules and modes of management; programme, to a co-ordinated set of activities of limited scale and budget; and project, to a non-divisible action with a mode of management, a schedule and a budget that are well defined from the outset (EC, 1999).

Project evaluation tends to focus on the assessment of the socio-economic viability of a distinct set of alternative options. The objective here is to determine the best option for achieving the programme or project alternatives or the most cost-efficient solution (Sugden and Williams, 1978; EC, 1995, 1997c).

Policy and programme evaluation are broader. According to Rossi and Freeman (1993) the objectives of policy and programme evaluation are primarily to provide input to decision-making by developing a rationale and justification for action. This definition restricts policy and programme evaluation to the agenda-setting phase. However this need not be the case. Pearce and Hett (1999) consider it possible to carry out a policy or programme evaluation also ex-post, for instance, however they underline that this is much more complex than at the project level.

The reasons they give are as follows:

- Projects have a well-defined project cycle which tends to be well understood. In contrast, policies are far more difficult to evaluate because their consequences are not well understood.
- Projects are small relative to the scale of the national economy, thus even when they fail they do not have major consequences for the population at large. The stakes are much higher in the case of policies. Next to having widespread impacts, they produce long-term effects which might be more important to assess than the direct short-term impacts.
- The quantification of impacts is likely to be more difficult for policies than for projects because of the plurality of intermediate variables which need not be specific to the policy under investigation.

On the same subject, Sansom *et al.* (1999) note the problematic nature of the timing of policy evaluation or impact assessment that derives from the mismatch between the desire for early policy evaluations of major interventions (for the purpose of guiding future interventions) and the time taken for the impacts to occur.

The Decision-Making Context

The political nature of evaluation (cf. Atkinson and Cope, 1994) is closely connected to the way it is situated in the decision-making context. This is characterised by a plurality of actors, ideas and interests. Each of the actors has their own agenda and own representation of the public interest. Reaching a decision is thus often a process of, or an attempt at, reaching consensus or compromise. Deliberation and bargaining are intrinsic parts of this process and evaluation can become the means to exert pressure on specific interest groups. This is an aspect of evaluation that one needs to be aware of. An understanding and knowledge of the decision-making context is important in order to avoid the abuse of one as an evaluator or of the evaluation results.

The European decision-making context is a particularly complex one, argue both *Francis McGowan* and *Ingmar von Homeyer* in their respective contributions on transport and environmental policies. EU enlargement is likely to increase this complexity, hence also the calls for institutional reforms. These are outlined and discussed by *Gerhard Rambow* also in this volume.

The term currently in use for describing the European decision-making context is multi-level governance. The goal of achieving conditions of 'fair competition' are at the core of harmonisation attempts in various policy sectors which, in turn, resulted in the emergence of supranational decision structures and in an impressive activity of the Union (previously the Community) in the regulatory field. Over 50 per cent – some argue closer to 80 per cent – of contemporary national legislation derives from the transposition of European directives and regulations entailed in the so-called *Acquis*. Several of these regulatory directives are in fact about deregulation: in transport, specifically, they are about removing barriers to market access and thus competition, and they are about liberalisation. A fair proportion deals with the establishment of 'product' standards: for instance technical standards for trucks, social standards about working conditions; as well as safety and environmental standards. In some fields, particularly fiscal harmonisation and pricing, harmonisation has not yet been achieved but is ongoing.

Across all modes, the harmonisation of legislation is however not equivalent to implementation. Indeed the implementation deficit, the spatial distribution of which is not uniform, represents a major constraining factor when establishing the 'background' conditions for evaluation.

In the field of infrastructure planning the European Union also has a major role to play, however its influence is constrained by the principle of subsidiarity (much in the same way that at the national level it is often argued that the competency for regional infrastructures lies with regional or local authorities). Insofar as many major infrastructure projects form part of networks that are transboundary, this raises the problem of co-ordination. No adequate solution has been found to this problem. This often makes for 'bizarre' situations where transboundary projects are evaluated differently by different national administrations, using different traffic forecasts and different methodologies.

Multi-level governance is also characterised by a more active involvement of economic interest groups as well as of social movements, like environmental groups, workers' associations or trade-unions.

How can evaluation take into account all these factors? Probably the most honest and sensible answer is that it *cannot*, especially considering the structure of commissioning evaluation, i.e. through specific agencies representing specific interests or having a specific stake in the decision process. Instead, Sen (1997) argues, it might be better to accept the differences across agencies, thus also evaluation studies, and be explicit about the so-called 'areas of control' of the evaluation. This then implies that the emphasis is placed less on harmonising evaluation methods, more on using the results in a constructive way in the decision process, re-shaped as a process of deliberation among relevant actors. Far from shifting the responsibility away from the evaluator this implies being stricter on the quality standards applied to the use of evaluation methods and tools.

Tools of Evaluation

There are several tools that can be used for evaluation. These can be classified mainly along four dimensions:

- The analytical framework from which they emerged and within which they mostly operate.
- Their suitability for policy, programme or project evaluation.
- Their suitability for different phases of policy analysis (i.e. ex-ante, monitoring or ex-post).
- The extent they rely on statistical or mathematical methods or tools or alternatively on 'softer' methods.

Existing classifications like those proposed by government bodies tend to consider primarily the second and third dimensions above, i.e. tools are classified according to their suitability for policy, programme or project evaluation and with reference to the timing of the evaluation exercise.

We would contend that if classification schemes are to be more than just as an inventory, they have to consider the two other dimensions mentioned above, namely the analytical framework of the method in question, and the extent of reliance on statistical or mathematical methods. We explain below why we think these two dimensions are equally, if not more, important for assessing the suitability and applicability of different methods to different problems encountered.

Clarifying Assumptions and Not Just Objectives

The analytical framework or meta-level of evaluation is not one which is today often dealt with explicitly. We agree with *Wayne Parsons* in this volume that this is the root of many flawed or unnecessary evaluations at either the project or programme level.

Let us take an example. Either it is taken for granted that cost-benefit analysis is rooted in welfare economics or this is assumed not to be important for either the demand or the supply side of evaluation. Yet cost-benefit analysis makes some very strong assumptions about social welfare comprising the sum of individual welfare and being a reflection of the aggregate sum of the individual's 'willingness to pay'. This is clearly one and not necessarily the only possible definition of what social welfare comprises. Evaluation strategies differ with respect to the extent to which they are driven by considerations about efficiency, economy and effectiveness, or considerations about equity and ethics.

Another difference concerns the treatment of causal relationships and quantified data. Mainstream approaches to evaluation, like neo-classical economics or public choice, insist on the importance of causal relationships that can be objectively verified and are keen on quantification or on tools that allow for quantification. Non-mainstream approaches, like interpretivism or critical realism, are more keen on recovering interactions and interrelations and insist less on quantification.

An interesting parallel is the debate within Operational Research/Management Science about 'soft' operational research (see, for example, Rosenhead, 1989). Here the argument has been that an increasingly formal and mathematicised style of analysis has taken the subject away from its roots as a multi-disciplinary, problem-centred approach to answering difficult and important questions about how organisations should tackle problems. Although strongly quantitative, operational research has undoubtedly had major successes and continues to be of value in the right context. However, the dominance of the quantitative aspect has threatened to marginalise operational research in debates about major strategic or social issues, because it was not well-suited to reflecting evaluations of competing alternatives. The soft operational research movement has sought to create a more balanced portfolio of approaches to problem-solving with more emphasis on process and facilitation of debate between stakeholders.

Several problems can arise by failing to be clear about the underlying assumptions of analytical frameworks used in evaluation. One frequent result is the subsequent failure to be clear about the objectives of evaluation exercises. This is a typical problem in multi-governance settings, like the European Union: the elaboration of policy programmes, and, subsequently, of evaluation tenders, represents itself a complex and often ill-structured process of bringing different and often contradictory objectives under the same hat. Another possible outcome is serious misunderstandings between those who commission and those who are assigned the task of evaluation.

Perhaps the most serious problem is the failure to understand or rightly interpret the results of an evaluation exercise and, in the case of a multi-tier evaluation exercise, to integrate the various results. Considering that policy analysis often necessitates such multi-tier evaluation exercises, it would seem that what often comes under the diagnosis of 'lack of co-ordination' is not simply a problem of organisation, but fundamentally also a problem in establishing a basis for 'reflective conversation'. The latter is only possible if at least a minimum level of common understanding is available.

Tools Are Not Just Tools

The second dimension that is often not given adequate consideration concerns the implications of using specific tools, in particular statistical methods, for the object of study. As *Frank Haight* in this volume notes, 'a statistical tool, like any tool, needs to be used carefully and wisely if it is to be effective'. There are two dimensions to this. The first is that a minimum and up-to-date knowledge of statistics is necessary if these are applied: Haight lists several examples which show that evaluation can go wrong due to the lack of statistical knowledge among those who apply it. Statistics is itself a science and in that relies on a set of assumptions. The evaluator needs to be aware of these prior to specifying an evaluation design.

The second dimension has to do with the 'choice among relevant assumptions, data and methods of analysis'. Haight argues that this choice has little to do with intuition or with theory. We would tend to agree that it has little to do with the theory of statistics, however, we would contend it has much to do with social theory. In order to be able to specify hypotheses and assumptions, it is important to have a certain understanding of the object under study. For policy-relevant issues, whether at the project or the programme level, this in turn requires a comprehensive knowledge of the policy in question, the decision-making context but also the role of policies as sets of interventions in the world of social relations.

The same arguments can be made also for tools like cost-benefit or multi-criteria analysis. In his contribution to this volume, *Michel Beuthe* traces the methodological development from cost-benefit (CBA) to social cost-benefit (SCBA) to multi-criteria (MCA) analyses (horizontally as well as vertically, i.e. within each methodological discipline) as a series of incremental attempts to overcome problems that are not solely theoretical but which additionally carry serious practical implications. Thus one major problem with cost-benefit analysis is its theoretical reliance on market values and by extension shadow prices; on the other hand, the more flexible design of multi-criteria analysis can increase the likelihood of double counting. Both methods face problems with the specification of weights to apply to different criteria, albeit in different ways: cost-benefit analysis in view of the difficulties involved in measuring reliably the 'willingness to pay'; multi-criteria analysis in adopting a 'subjectivist' approach to this and relying on the decision-maker or a round of experts to determine how important any particular type of good or impact is for social welfare.

The requirement to define and account for sustainability criteria in cost-benefit analysis has accentuated calls for the latter's modification but, at the same time, it has revealed the theoretical and empirical constraints. *Klaus Rennings* and *Sigurd Weinreich* show in their contribution that whilst it is possible to modify cost-benefit analysis to take environmental impacts into account, this can probably not be reasonably done without giving up on the sole use of monetary valuation techniques or without accepting the use of multiple monetary values.

Following Haight and Parsons, this, in our opinion, raises the question of whether it is not wiser to *not* modify cost-benefit analysis thus keeping its internal consistency and coherence. Instead one ought to more carefully reflect the use and application of cost-benefit analysis.

Towards Harmonisation or Flexibility?

Is it at all possible to develop one universal evaluation method or evaluation framework for transport that satisfies the various user needs as well as the objective demands of policy? An equivalently important question is, would this be desirable even if possible?

Parsons claims that an integration is possible and desirable if understood as a process of clarification which, in turn, is important for 'reflective' communication and possibly consensus-building. It would be a mistake to equate integration with the achievement of a value-free, or worse, value-uniform policy; or to couple this with the denial of conflict: policies, however well designed, are unlikely to just have winners.

Instead our attention ought to focus on developing (and subsequently applying) generic guidelines for policy and project evaluation that help establish standards or conventions about what is a good and professional evaluation and how specific methods are to be used. Lack of integration can either be substantive or conjured in nature. In the former case the reasons for failing to reflectively communicate and arrive at a compromise or consensus in decision-making are for better or for worse unrelated to evaluation – they are instead the result of a conflict of interests or of visions. This, we would contend, is ultimately less serious than communicative failure resulting from bad evaluation practice or the naive application of scientific methods.

Integration understood as aiming towards rational and analytical thinking and co-operation is a worthwhile goal to strive for – in policy as much as with regard to evaluation. Knowledge and proficiency are important background conditions in this respect. Integration, however, is not conditional on harmonisation and the levelling off of differences in approach. Indeed flexibility is instead called for. Because only through the use of several and of different approaches will it be possible to capture and understand system complexity.

References

Atkinson, R. and Cope, S. (1994), 'Urban Policy Evaluation: Science or Art?', Paper presented at the ESRC Research Seminar on Urban Policy Evaluation, Cardiff.

Banister, D. and Berechman, J. (eds) (1993), *Transport in a Unified Europe*, North-Holland, Amsterdam.

Banister, D. and Lichfield, N. (1995), 'The Key Issues in Transport and Urban Development', in Banister, D. (ed.), *Transport and Urban Development*, London, E&FN Spon.

Chelimsky, E. and Shahdish, W.R. (eds) (1997), *Evaluation for the 21st Century*, SAGE London.

European Commission (1994), *Strategic Environmental Assessment – Existing Methodology*, EC, Brussels.

European Commission and Transport Evaluation Group (1995), *Framework for Transport Investment Evaluation*, EC, Brussels.

European Commission (1997a), *Case Studies on Strategic Environmental Assessment; Final Report: Volumes 1 & 2*, EC, Brussels.

European Commission, DGXIX (1997b), *Good Practice Evaluation Guidelines Handbook*, EC, Brussels.

European Commission (1997c), *Financial and Economic Analysis of Development Projects*, EC, Luxembourg.

European Commission (1999), *MEANS collection – Evaluating Socio-economic Programmes*, EC, Brussels.

Freeman, H.E., Rossi, P.H. and Wright, S.R. (1980), *Doing Evaluations*, OECD, Paris.

Giorgi, L. and Pohoryles, R. (1998), *TENASSESS Deliverable 5, Interconnections Among Tasks: A Guide to the Fourth Framework Strategic Transport Research Programme*, ICCR/EC, Vienna/Brussels.

Hoon Oum, T. *et al.* (eds) (1997), *Transport Economics; Selected Readings*, Harwood Academic Publishers.

Layard, R. and Glaister, S. (ed.) (1996), *Cost-Benefit Analysis*, Cambridge University Press, Cambridge.

Leleur, S. (1995), *Road Infrastructure Planning*, Denmark, Polyteknisk.

Palumbo, D. J. (ed.) (1987), *The Politics of Program Evaluation*, Sage, Newbury Park.

Parsons, W. (1995), *Public Policy: An Introduction to the Theory and Practice of Policy Analysis*, Edward Elgar, Aldershot.

Pearce, D., Hett, T. *et al.* (1999), *Review of Technical Guidance on Environmental Appraisal*, UK Department of the Environment, Transport and Regions (DETR), London.

Rosenhead, J. V. (ed.) (1989), *Rational Analysis for a Problematic World*, Oxford, Wiley.

Rossi, P.H and Freeman, H.E. (1993), *Evaluation: A Systematic Approach*, SAGE, California.

Rus, G. de, Nach, C. (eds) (1996), *Recent Developments in Transport Economics*, Avebury, Hants.

Sansom, T., Pearman, A.D., Matthews, B. and Nellthorp, J. (1999), *The SITPRO Methodology*, Deliverable 2, SITPRO (Study of the Impacts of the Transport RTD Programme), EC, Brussels.

Sen, A.K. (1997), 'Shadow Prices and Markets', in Layard, R. and Glaister, S. (eds), *Cost-Benefit Analysis*, Cambridge University Press, Cambridge.

Sugden, R. and Williams, A. (1978), *The Principles of Practical Cost-Benefit Analysis*, Oxford University Press, Oxford.

Turro, M. (1999), *Going Trans-European; Planning and Financing a Transport Network for Europe*, Pergamon, Amsterdam.

US General Accounting Office (1991), *Designing Evaluations*, Program Evaluation and Methodology Division, US.

2 The Policy-Making Process in the European Union

FRANCIS McGOWAN

Introduction

This chapter provides an overall perspective on the EU policy making process as background to the evaluation of EU transport policy. 'Policy-making' is not normally a phrase used to describe the functioning of international agreements, however; it implies a polity or political system of the sort which operates in a domestic political environment. Yet the term is appropriate here because, along a spectrum between national political systems and international organisations, the EU is closer to the former than to the latter. It shares many similarities with institutions, practices and problems in national policy settings (Hix, 1995, 1999) and, accordingly, we are able to consider EU policy processes from the vantage point of debates within the political science and policy analysis literature which assume the state as their focal point. That said, however, we also need to recognise that the EU system is rather distinctive due to the 'international' aspects of its operations and the institutional bargains which have shaped its development. Factors such as the incomplete nature of the political system, the multilevel nature of policy making (and of influencing policy) and the relatively high propensity towards regulation as a mode of governance render the EU unique (Laffan, 1998) and we need to take these distinctive characteristics into account. As a result the paper also draws upon concepts and approaches which have been developed within the academic community specifically to explain how the EU works.

The chapter has four sections. The first reviews the historical evolution and institutional development of the EU, providing a context to the rest of this chapter and the book. The second elaborates upon some of the distinctive features of the EU considers how the EU has been analysed over the last fifty years. The third section looks at the institutional interactions within the EU both at the European level and between the EU and the national (and other) levels. The fourth section looks at how we can

14

understand policy change within the EU. In the conclusion we consider how the EU policy process can be understood in terms of some of the principal categories of policy making. The paper draws upon a range of sectoral examples with particular attention given to the energy and transport sectors and their interaction with market liberalisation, environmental protection and regional development policies.

The EU in Historical/Institutional Context

Policy-making (particularly when understood as problem-solving) is usually understood as a discrete process, divorced from history and politics. Often, that process is systematised as a series of steps amenable to technocratic analysis: from agenda setting to evaluation, policy-making is interpreted as a mechanical enterprise, administered – and explained – in a rational fashion. Whether or not this approach provides anything more than a heuristic, it is open to challenge in an EU setting: in this case, it is impossible to understand the particular nature of the policy process without some reference to the historical roots of European integration and the institutional structures which were created. Table 1 outlines the evolution of the European Union (as well as an indication of the parallel development of EU policies in the transport field).

The EU's roots lie in the attempt by the French and German governments to deepen the post war rapprochement between their countries (Dinan, 1999). The European Coal and Steel Community (ECSC) was created in order to integrate markets for what were then strategic resources in the post war economy, thereby overcoming a historic source of tension (Diebold, 1959; Gillingham, 1991; Spierenburg and Poidevin, 1994). In its ambitions and processes, the idea of such integration harked back to earlier ideas of a link between economic and political integration and – ironically – of the scope for rational problem-solving amongst experts (Radaelli, 2000; Spinelli, 1970). The ECSC brought to the fore the idea of creating a common market in Europe (though as much by planning mechanisms as by trade liberalisation) backed up by an independent and powerful secretariat (the High Authority, later to become the Commission) and a strong system of law. These latter features, followed up in the European Economic Community (EEC) and – less significantly, Euratom – came to define the 'supranational' nature of European integration: member states 'pool'[1] sovereignty in order to obtain the benefits of integration. At the same time however the fact that these arrangements were agreed by sovereign governments and that those governments brought their own concerns and preferences to the negotiating table and to the workings of the new Communities meant that the new system was framed (and constrained) by national interests. This tension between supranational and intergovernmental aspects is reflected in the original institutional structure of the EU, the elements of which have survived even if their relative balance between them may have changed over time.

The Treaty of Rome, which created the European Economic Community (EEC), built upon the structure set out in the ECSC Treaty by allocating key roles to four institutions:

- A Council of Ministers which was to negotiate and agree on the basis of Commission proposals. It was to have the final say in EU decision-making. The Council's work is sectorally defined and is underpinned by an extensive system of committees and by a Committee of Permanent Representatives.
- A Commission which was to have the sole power of initiative in framing policy and which was intended as an independent bureaucracy pursuing the Community interest. It comprised an international secretariat, sectorally organised, and a set of political appointees from each member state who were to provide leadership for the Community and pursue its policy objectives.
- A Court of Justice which was to rule on disputes between member states and the Community institutions. As with the Commissioners, the judges were appointed by member state governments but were expected to act in the Community interest rather than on national lines.
- An Assembly (later renamed the European Parliament) which was to offer opinions on Commission proposals. This, too, was appointed for many years though direct election was envisaged and eventually transpired in 1979.[2]

The evolution of the Community since its founding has been marked by a tension between supranational and intergovernmental tendencies, most obviously between the Commission and Council of Ministers and manifest in political milestones. Underlying these apparent oscillations, however, has been a high degree of continuity in the making of policy, thereby giving substance to claims that a 'EU system' has emerged.[3]

Table 2.1 EU Evolution and Transport Policy Development

EU Evolution	EU Transport Policy Development
1950s Beginnings	**1950s Transport and the Treaties**
1950 Schuman Plan	
1952 Treaty of Paris – ECSC	1952 Treaty of Paris: liberalising transport of coal & steel
1957 Treaties of Rome – EEC/EAEC	
1960s Consolidation and Crisis	1957 Treaty of Rome – Common Transport Policy
1962 Common Policies	**1960s A Slow Start**
1965 Luxembourg Crisis	1961 Memorandum on a Common Transport Policy
1966 Luxembourg Compromise	
1967 Merger Treaty	1962 Action Programme on a Common Transport Policy:
1968 Customs Union completed	
1970s Enlargement and Recession	1967 'Relaunch' of CTP
1970 Werner Report on Monetary Union	**1970s Limited Progress**
1973 Accession of Denmark, Eire and UK	1979 Liner Conference Regulation
1973 Oil crisis (and economic recession)	1979 First Memorandum on Transport Policy
1974 Budgetary expansion	
1979 EMS agreed	**1980s Recovery and Liberalisation**
1979 Elections to European Parliament	1980 Commission proposal on CTP priorities
1980s Eurosclerosis and Europhoria	
1981 Accession of Greece	1983 EP takes Council of Ministers to Court over CTP
1984 Fontainebleau Summit	
1985 Internal Market White Paper	1983-5 New initiatives in inland, air and maritime transport
1985 Single European Act	
1986 Accession of Portugal and Spain	1985 Internal Market White Paper – Transport Services a Priority
1989 Social Chapter Agreed	1989 TENs proposed
1990s A European Union?	**1990s Redefining the CTP**
1990 IGCs on EPU and EMU begin	1991 Railway liberalisation (91/440)
1991 Maastricht Treaty agreed	1992 Incorporating the Environment
1992 European Economic Area	1993-5 TENs in practice: financial instruments, priority projects
1995 Accession of Austria, Finland and Sweden	
1997 Amsterdam Treaty agreed	1995 Addressing Passenger Transport: The Citizens' Network
1999 EMU enters into force	1997 Infrastructure Pricing Proposals
1999 Commission resigns	1998 Sustainable Mobility
1999 Enlargement negotiations begin	1999 Railway Liberalisation
2000 Nice IGC	2000 Air Transport Passenger Rights

The European Community of the 1960s made some progress in obtaining its initial objective – a customs union – as well as developing its competences. A steady flow of legislative proposals and new initiatives put flesh on the Treaty commitments and broadened the Community's agenda. Another important development – although not necessarily apparent at the time – was a series of judgements of the European Court which asserted the primacy of European law over national law (Mancini, 1989; Burley and Mattli, 1993; Garrett, 1995; Alter, 1998). However it would be wrong to suggest that the EC was inexorably in the ascendant and national authorities on the defensive. The 1960s also saw national governments assert their sovereignty. The 'empty chair' crisis of 1965 – when the French government effectively boycotted the Community – and its resolution in the form of the 'Luxembourg Compromise' demonstrated that member states were willing and able to veto proposals they judged to be against their national interest, bringing about a de facto system of unanimity in most areas (Hoffman, 1966; Dinan, 1999).

The 1970s were to see a consolidation of this tendency with enlargement and economic crisis compounding the problem (Keohane and Hoffman, 1992; Tsoukalis, 1997). Enlargement effectively widened the range of interests to be resolved and the UK's accession presented particular difficulties which were to impede integration for at least a decade. Economic crisis prompted most governments to retreat to the national level to solve restructuring problems responding at best lukewarmly (often with some hostility) to attempts to develop European responses. Nonetheless the period saw some deepening of EU activities with the development of new policy responsibilities. The 1970s saw progress in the development of a European regional policy (in the process enhancing the budget), the first steps towards monetary union and, in a detached way, measures to enhance foreign policy cooperation. The first direct elections to the European Parliament were also held, strengthening its claim for more powers (Westlake, 1994; Earnshaw and Judge, 1995).

It was not until the mid 1980s that a serious revival in the dynamic of European integration took place. Whether this was because of a recognition by key governments that deeper integration was needed (Moravscik, 1991) or because of the pressure of European business or the appointment of a more proactive Commission (Sandholtz and Zysman, 1989), the pursuit of the European project appeared to take on a new urgency. A commitment to a single market not only sought to streamline the European economy but seemed to put a greater emphasis on market liberalisation and the use of EU powers (particularly the competition rules) to encourage that trend

(Commission of the European Communities, 1985; Tsoukalis, 1997). The Single European Act (effectively a series of amendments to the Rome Treaty) not only changed the institutional balance of power (by eroding the role of unanimity in the Council of Ministers and granting limited decision making powers to the European Parliament) but also extended policy competences further (notably in the fields of environmental and research policy[4]). Moreover this dynamic seemed relatively unaffected by a further wave of enlargement to include Southern European states (although the apparent ease of their accession – relative to the 1973 enlargement – was prefaced by tough negotiations and underpinned by the allocation of significant funds to assist their development).

The late 1980s appeared to be a period where political leadership, public support and historical circumstance converged to push forward the idea of integration. Sensing the momentum, the Commission President Jacques Delors sought to push forward the integration project, claiming that approximately 80 per cent of legislation would be determined at the EU level (Richardson, 1996; Ross, 1995). The fall of the Berlin Wall and changing geopolitics of Europe also contributed to efforts to deepen integration. The IGCs on Economic and Monetary Union and on Political Union offered an opportunity to push forward, and while there were some key disagreements and even some opt outs from the commitments made, the Maastricht Treaty fulfilled these ambitions, setting out an agenda for developing European roles in foreign policy and home affairs as well as achieving an Economic Union (though in each case removing the policies from the traditional community arrangements, retaining a substantial role for governments in the definition and development of policy). In addition the scope of EU policies was broadened along with an intensification of 'supranational' mechanisms (through more qualified majority voting – QMV – and an enhanced decision-making role for the EP). The momentum of integration, however, was to come unstuck as popular unease over the pace of reform manifest itself in referendum results in Denmark and France (Dinan, 1999).

Thus, although Maastricht locked member states into a further IGC, the reaction to that agreement was such that further radical reform would be politically impossible. Moreover, while the mid 1990s saw another relatively smooth accession, of EFTA states,[5] as well as progress along a number of policy fronts (notably telecommunications liberalisation and social and environmental protection), when it came to the 1997 IGC, the results at Amsterdam were rather limited, extending competences in a few areas (notably with regard to the role of the European Parliament), developing the 'intergovernmental pillars' of justice and home affairs and a common foreign and security policy (bringing them slightly more into the orbit of the established European institutions), and elaborating on the concept of Europe citizenship. However the IGC failed to address the most pressing problem, the institutional details of how the EU would function after the next wave of enlargement (Junge, 1999). A further IGC was convened, the results of which were agreed at the Nice Summit in December 2000.

The period between the two IGCs was to be one of the most difficult for the EU. Not only was the most ambitious and difficult enlargement (involving 12 states mostly from Central and Eastern Europe) under negotiation, but the legitimacy of the EU itself was under increased pressure. The resignation of the European Commission following a series of investigations into misconduct, falling participation rates in the European elections, a crisis in relations between Austria and the rest of the EU and a difficult birth for the Euro cast doubt on the degree of effectiveness of, and support for, European integration. The debates on the IGC were paralleled by a series of internal reforms, most notably in the Commission where a set of administrative changes have been proposed which are designed to improve the efficiency of decision making (European Commission, 2000b).[6]

The results of the 2000 IGC proved to be highly controversial. While the Nice summit saw the launch of many potentially highly significant initiatives (including progress on European defence policy and the signing of the Charter of Fundamental Rights), the central question of institutional reform was not fully addressed. While there was an agreement on extending the application of QMV and reforms designed to limit the number of Commissioners (removing the larger states' second Commissioner and capping the size of the College at 27), the question of the weighting of votes within the Council of Ministers proved much more controversial. While the reforms meant that the voting power of larger states was more in line with their populations (traditionally the system was skewed to favour smaller member states) this shift was heavily criticised as concentrating power in the hands of the largest states. Indeed for some the reforms were seen as shifting power back to the member states. Moreover, the debates around the IGC revealed very different visions of the EU amongst the member state governments and the European institutions.[7]

Notwithstanding these difficulties, the long-run evolution of the EU has been remarkable. As Table 2 shows, the scale of EU activity has increased dramatically in terms of administrative load. While there is some evidence that the system has plateaued (and the Commission itself has signalled that it has to limit its activities in the absence of extra resources) the level of activity is still substantial and covers most of the key areas of modern governance. While the future trajectory of the EU is uncertain, its continued presence as a major focus for policy co-operation amongst the states of Western (and – in time – Eastern) Europe is not in question.

Table 2.2 Growing Density of the EU as a System of Governance

	1960	1975	1990	1994
Council Decisions	10	575	618	468
Commission Decisions	6	838	1367	2461
Council Compositions	7	12	22	21
Council Working Groups (WG)	10	91	224	263
Council Sessions	44	67	138	98
Coreper + WG Sessions	612	2215	2128	2789
A-Grade Officials in Commission	632	2087	3642	4682

Source: Wessels/Laffan

The EU Polity and Policy Making

We noted in the introduction that, in a number of important respects, the EU was closer to national polities than to international organisations. This judgement is based on the development of effective institutions, accepted procedures and extensive policies which mark it out from most other regional integration experiments.[8] Yet it falls well short of being like a nation state. Indeed, it could be argued that, while the EU constrains its member states in various ways, it is itself constrained by those same states, potentially leaving a gap between the two levels of governance. This does not mean that we should treat it as unique model which is not amenable to the techniques, metaphors and concepts normally applied to national systems but that we should qualify our analysis by taking account of its relative distinctiveness as a system of governance. That distinctiveness has a number of characteristics.

The EU is an 'Incomplete Polity'

This characterisation highlights the contrast between on the one hand, unusually dense institutional structures along with substantial powers and responsibilities in some areas, and on the other a range of less developed – or even excluded – policy areas. Institutionally the EU has a very well developed bureaucracy and legal system as well as a legislative wing which has accumulated powers over the last two decades. In terms of policy there are several aspects of economic governance – and increasingly other areas – where EU level decisions have both prevailed over national rules and become the primary target for lobbying. Yet there is no 'EU government' (indeed the size of the Community institutions is relatively small by comparison with most governmental bureaucracies), the budget of the EU is extremely small (less than 2 per cent of EU GDP compared with an average 40+ per cent in most member states) and some policy realms are largely absent from the EU agenda (notably health and education policies, though it is true that even here, the EU has an indirect impact and there are attempts to develop an explicit EU dimension). In both institutional and policy terms, however, the EU depends upon the willing delegation of powers from – and subsequent implementation of legislation by – member states. Governments remain reluctant to transfer sovereignty in some areas and in a few cases appear less than enthusiastic about following through on policies even though an agreement has apparently been reached (Richardson, 1996). This incomplete nature of the EU polity has a direct impact upon policy-making. In some cases the ramifications of an EU policy risk falling beyond its competence – the acts of EU institutions may impinge on areas where they do not have responsibility. In other cases, however, there may be enforcement problems given the lack of resources at the EU level (the general view of the Commission and others is that it is overstretched). In both cases the shortcomings may pose credibility or legitimacy problems which could serve as a constraint upon future conduct. As such, they raise questions about the role of 'politics' in EU policy-making, particularly the so-called 'democratic deficit', though this may change as the role of the European Parliament evolves (Neunreither, 1994).

Propensity Towards Regulation

The incompleteness of the EU polity is also reflected in the instruments which it is able to utilise and roles it can play within the policy-making process. Governments have traditionally pursued policy goals by a variety of mechanisms such as provision with owning, providing, purchasing, promoting, (re)distributing and regulating (Grant, 1993; McGowan and Wallace, 1996). Which of these policy modes are available to the EU institutions? The EU is scarcely able to make policy through 'ownership'. The highly bureaucratic processes of purchasing which the EU has imposed upon itself scarcely lend themselves to anything but the bluntest policy steering, though attempts to inform purchasing decisions or the allocation of Community funds by broader Community goals, such as economic and social cohesion or environmental protection, may have some impact (EC, 1999). The Commission has a role as a promoter, through the allocation of funds and attempts to develop information campaigns and encourage bench-marking and networking, but the scale of the EU budget limits the scope of such policies (Laffan, O'Donnel and Smith, 2000). Funding constraints also limit its direct distributive role, though for some regions the impact of EU funds is significant (Allen, 2000).

What *can* be done at the EU level is regulation, setting rules and monitoring their compliance thereby constraining or encouraging governments, firms and other actors. The combination of a strong legal system and a relatively effective administration positioned at the hub of a series of policy networks means that the Commission, backed up by the Court, has been able to act as a powerful regulator in a range of policy areas, developing a greater influence over policy than was the case in the past (Gatsios and Seabright, 1989; Majone, 1994; McGowan and Wallace, 1996). Although regulation appears to describe a relatively narrow range of activities and outcomes, in practice a wide variety of objectives can be pursued through this means, often obtaining similar results to those pursued by other modes of governance. The most obvious example is that of distribution: establishing or amending a series of rules on matters such as market structure, working conditions or environmental emissions will have distributive effects (intentional or otherwise).

Figure 1 illustrates the ways in which the EU replicates the roles of the state according to three categories: regulatory, developmental (equivalent to state intervention to foster economic development through direct support) and distributive (a variety of compensatory mechanisms either direct or indirect in character). As the figure shows, the strongest EU policies are those relating to the regulatory realm.

The importance of the regulatory dimension in EU governance goes beyond the simple fact that the institutional characteristics of the EU make it compatible with, and capable of, regulation. It is also important because increasingly national authorities are themselves obliged to adopt regulation as a mode of governance, given the constraints that they face in terms of expanding budgets, the orthodoxy against interventionist economic policies and the opportunities which arise in terms of pursuing apparently costless social policies (shifting the burden of compliance to others) and enhancing competitiveness (either 'deregulation' or 'technology forcing'). Yet they have to do so generally in cooperation with their EU partners and with the EU authorities. In a sense therefore the EU acts as both a constraint and a collaborator on national regulation as well as a regulator in its own right (though the latter may itself be constrained by the preferences of member states).

Figure 2.1   State Roles and EU Competences

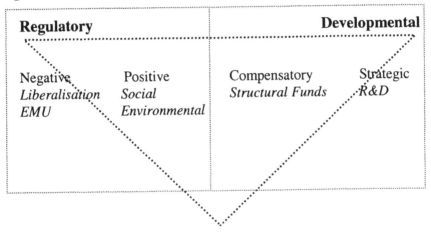

The emphasis on regulation as a mode of governance – with its connotations of independent and rational decision making – also highlights the importance of expertise in the making of European policy. The image of the European Commission as a 'technocratic' body has an element of truth to it while the bulk of decisions taken by the European institutions are 'low politics', technical questions largely resolved in specialised working committees where expert opinion and knowledge is at a premium (Radaelli, 2000; Cini, 1997). That is not to say, however, that such regulatory decisions are apolitical; on the contrary, different ideas and technical solutions may reflect particular clusters of interests within industries or member states. Indeed, one characteristic of EU regulation is that it can be a conduit for particular states or clusters of states to impose their particular regulations upon the EU as a whole. There is, in other words, no single EU regulatory model but a patchwork, reflecting the diversity of approaches and interests in the EU (Heritier, 1996).

Multilevel Policy Making

Nor is it only the mode of governance which marks out the EU. The EU pattern of policy making is of necessity multi-tiered, sharing in that respect some similarities to federal systems (Marks *et al*, 1996; Scharpf, 1988). Moreover, the multiple tiers (national-EU, increasingly local and global as well) are overlaid by multiple points of access at each level. Not only does this make for a highly complex system of policy making but it also provides interests not only with a variety of conduits to influence policy but also the possibility for a variety of coalitions which are dependent upon the point of contact (Grande, 1996).

One effect may be to 'squeeze' the nation state by permitting alliances between the subnational and supranational levels at the expense of national authorities.[9] But such arguments need to be treated with caution since they tend to be voiced in particular areas where issues, interests and resources are interacting (such as in the area of regional policy). Elsewhere it is not obvious that such a 'squeeze' is taking place. Overall, however, a more modest process of multilevel governance appears underway in a wider range of policy-making, raising the question of subsidiarity – at what level is policy best carried out (McGowan and Seabright, 1995). In any case such interactions only add to the complexity of policy-making within the EU.

Taken together, these characteristics have tended to shape a somewhat distinctive approach to policy making within the EU. However, there may be different emphases, dynamics and styles according to the type of policy that is being made and the sector involved. Moreover it is clear that the interplay between these factors and different realms of policy-making does vary over time and sector. Thus in the energy and transport sectors, the record of EU policy making has been mixed, though more recently it has become more effective, particularly where regulatory modes of governance have been deployed. However, whatever the specificities of particular sectoral policies, or even particular policy decisions, the changes do reflect a broader pattern of EU policy making. On that basis we can examine in more detail the nature of that policy process.

Box 1   Approaches to the Study of the EU

In order to explain the EU policy making process it is worth considering how the EU itself has been studied. Until relatively recently (and with some important exceptions) the core disciplines analysing the EU were economics, international relations and law, each of which focused on rather different aspects of the process.

The primary concern of economic analysis related to the trade effects of customs unions (whether the EU was trade creating or trade diverting) (Machlup, 1977). As the EU's own policy competences developed however, a broader range of analyses (rooted in industrial economics and organisation) emerged along with various sectoral specialisms, not least in the area of transport economics (Pelkmans, 1980; Gwilliam and Mackie, 1975). Almost without exception, the concern of this strand of analysis was the effects of policies rather than the policy process. More recently there has been some attempts to apply public choice and game theory approaches to institutional design.

International Relations debates have tended to revolve around the issue of the dynamic of integration itself. Drawing upon older, and idealist, traditions of international relations relating to international cooperation, the initial concern was to analyse how economic integration might contribute to regional peace. The so-called neofunctionalist school predicted a broadly positive relationship between these on the back of a steadily increasing degree of integration and pooling of sovereignty, the latter facilitated by a virtuous circle of transferred loyalties and expanding tasks ('spill-over') (Haas, 1963; Lindberg, 1970). However as the hopes of the most ardent Europeans were apparently dashed, a 'realist' backlash determined that power politics and national interests were still the primary forces within Europe and that these would contain the process of integration (Hoffman, 1966). Events in the 1980s offered a revival of the theoretical debate (Zysman and Sandholz, 1989; Moravscik, 1991).

The profile of legal analysis in the study of the EU is largely a function of the importance of law in the process of European integration. Given the rules-based orientation of the EU, it is scarcely surprising that lawyers should seek to understand the workings of the new institution (Wyatt and Dashwood, 1987).

Since the 1980s, and the apparent revival of the EU, interest in explaining its operations has percolated down to a variety of other disciplines. Since that 'revival' built upon and endorsed a pattern of relationships and institutional arrangements, this academic interest reflected a sense of the EU as an established system of governance. In this period a number of other disciplines (sociology, anthropology) have tackled aspects of the EU. However, for our purposes the most significant development has been the afore-mentioned willingness to view the EU as a polity or system, from a political science perspective (Hix, 1999), or a process, from a policy analysis perspective (Richardson, 1996).

Institutional Interactions

The first aspect to examine is that of the interaction between institutions. By this we do not simply mean the relationships between the key organs of the EU (including member state governments) but also the interplay between them and other actors (interest groups, local government, governments outside the EU) as well as *within* the institutions themselves. Moreover we assess the effects of this network of interactions: has it led to substantial convergence across the EU (the so-called Europeanisation hypothesis) or has it been less effective, leaving member state governments and other actors able to circumvent decisions?

Inter-Institutional Interactions

Here we build on our earlier account of the particular roles of the EU institutions by focusing on and identifying the general patterns of interaction between them. The central dynamic in EU policy-making is generally perceived as that between the European Commission and the member states as represented in the Council of Ministers. In some accounts (and in some political debates) this relationship is described as conflictual and zero-sum, with member states defending their sovereignty over policy against encroachment from attempted supranational 'incursions' (Moravcsik, 1999). For others, policy-making is circumscribed by the institutional mechanisms for decision-taking (particularly the reliance on unanimity or super majorities) which, in the form of 'joint decision traps' (Scharpf, 1988) or 'log-jams' (Heritier, 1999) create substantial bottlenecks in policy-making. While there is some substance to such accounts, in practice the relationship is much more co-operative, reflecting the interdependence between institutions and the underlying commitment of most of the actors to the general goal of integration (even if they disagree on important details and, sometimes, on the final destination of the European project) (Wallace, 2000). The afore-mentioned institutional obstacles are to some extent overcome by a more informal approach embracing a mixture of 'escape routes', unofficial practices and other procedures which keep the process of policy-making ticking over (Heritier, 1999).

The source of this pragmatic praxis is to be found in the mechanics of EU policy making itself and in particular the system of repeated contacts between officials in the Commission and in government representatives (whether in Brussels or national capitals); the metaphor of 'engrenage' conveys a sense of the intense interaction between officials at all levels (Wallace and Wallace, 2000). In essence the EU policy process can be seen as rooted in 'communities' of policy makers working together, whether in working committees at relatively low levels of seniority (where many EU policy decisions are taken) or in the highly communautaire and consensus-driven setting of Coreper, the body which lies at the peak of a multitude of specialist committees and meetings and which brings together the national 'ambassadors' to the EU to prepare the ground for Council of Ministers meetings (where only the most contentious issues are debated) (Lewis, 1998).

Between the Commission and the Council of Ministers, there appears to have been some shift in the balance of power. The Commission has tended to be by definition the agenda setter and as such has been relatively entrepreneurial in pushing policies forward (Cini, 1997). However, there appears to have been a 'relative decline' in Commission activism in recent years and the evidence that increasingly its 'priorities' are responses to internal pressures or to 'external shocks' (Laffan, 1997). The interaction between the Commission and member states has become as much one of monitoring and liasing with national governments as that of seeking to expand the competence of the EU and the powers of the Commission itself. In contrast, the Council of Ministers (from the network of committees and working groups through to the European Council meetings of national leaders), not only remains the venue where national interests can be articulated and defended, but also appears to be the focal point for decision – making in those new competences which do emerge (Hayes, Renshaw and Wallace, 1996).

This role has been – and will increasingly be – constrained by the European Parliament, whose growing powers have complicated the legislative dynamic in the EU. Prior to the Single Act it had been a relatively powerless body aside from occasionally asserting its powers in regard to budgetary matters. Since the development of co-operation and co-decision procedures the Parliament has become a co-legislator with the Council on a very wide range of policy areas (Tsebelis, 1994; Earnshaw and Judge, 1995). However, these powers are constrained: just as decisions in the Council of Ministers require 'qualified' (i.e., large) majorities to pass (if a vote is taken) so the European Parliament needs to gather an absolute majority of its members for a particular decision to impose its will (Hix, 1999).

The Court is detached from these policy interactions (though it can and has been brought in where one institution has challenged the conduct of another). However that is not to relegate its role to adjudicating on the fine detail of EU law. On the contrary, its decisions frame the legal context within which European integration develops on a day to day basis as well as determining the limits of institutional power. Of particular interest is the broadly positive and reinforcing relationship between it and national courts (Weiler, 1994; Alter, 1998).

Box 2   Institutional Interactions and Common Policies – Exceptions to the Consensus Rule?

To a large extent it is the limitations rather than the potential of the EU policy process which are illustrated by the historical evolution of EU energy and transport policies. In both realms the Commission has displayed a high degree of activism both in proposing specific initiatives and in attempting to create an overarching framework for policy. The first attempts to develop common policies emerged in the 1950s for both energy and transport (Abbati, 1987, Alting von Geusau, 1975). In the case of energy this initiative also illustrates the Commission's 'entrepreneurialism' as there was no specific treaty base. However, this activism was not reflected in member state acquiescence. On the contrary the cases of energy and transport policy also provide numerous examples of how policy making in the EU can become deadlocked. Thus, during the 1960s and 70s many of the Commission's proposals on energy and transport policy were watered down, delayed or in some cases ignored (McGowan, 1998a, 1998b; CEC, 1992, 1994). In the case of transport this problem became so acute that the EP took the Council to the Court which found that it had failed to act on its commitment to develop a Common Transport Policy.

In part these problems were seen as symptomatic of the broader problem of achieving unanimity within the Council. However, it is not clear that voting reforms have necessarily helped: the proposal to liberalise electricity markets – subject to QMV rules – took nearly ten years from original proposals to finalisation and it seems that a similar timetable is in train for rail freight liberalisation. More importantly, perhaps, the pace of decision making may reflect the continued determination of at least some member states to retain autonomy in what was perceived to be a vital national interest and where strong patterns of self regulation within the sectors involved were already in existence.

The potential of the Court has manifest itself on a number of occasions in the development of energy and transport policy. In addition to the case noted above, Court rulings were instrumental in pushing forward the Commission's proposals to liberalise air transport markets in the 1980s (Argyris, 1989). The Court's role is likely to become more important as part of the EU regulatory process in applying EU law, particularly competition policy. As energy and transport markets shift from being 'exceptions' to being closer to other market sectors, the Commission and Court are likely to act as 'regulators of last resort' in a number of cases (EC, 2000).

'Intra'-Interactions

So far we have concentrated on interactions between institutions. However it is clear that the internal politics of Commission decision-making are critical to the setting of initial proposals. Moreover, negotiations within the Council of Ministers (among member states), within the Parliament (among party groups and other clusters) will be primary determinants of the policy outcome.

Within the Commission there are three key interactions, amongst the services (the directorates general and other parts of the administration), within the College and between the College and the services. In each case a key role is played by the individual cabinets of the Commissioners while the Commission secretariat is also important in co-ordinating the interplay between College and services and amongst the services themselves (Dinan, 1999). The inter-service debates are in many ways the most important as these highlight differences in perspectives between different directorates (Cini, 1997). The intra-College debates are critical in framing the final shape of a Commission proposal and reflect not only the responsibilities of the individual Commissioners but also, in some cases, national concerns (notwithstanding official rules) and political sensitivities about what is feasible (Ross, 1995).

As noted earlier, negotiations between the member states are driven by consensus and by the fact that EU decision making is an iterative process, locking officials into set routines and bargaining across issues, thereby offering considerable scope for compromise (Pierson, 1996; Heritier, 1999). However, national positions will also reflect the configurations of power and influence within each member state and this may make compromise more difficult on controversial issues. As noted earlier the processes of decision making normally ensure that agreement is largely reached at the committee or permanent representative level. Where the Council of Ministers itself becomes involved, the issues are usually more sensitive and difficult to resolve (Hix, 1999).

Box 3 Intra-Institutional Dynamics and Tensions

Within the Commission, the main tension has been between the sectoral (or 'producer') directorates on the one hand and cross-cutting (or 'horizontal') directorates on the other, where the general pattern has been that those directorates with a strong overarching ethos have sought to shape policy in those directions, while producer directorates have been more sensitive to the established arguments and interests within their respective sectors. Thus the 1989 paper on energy and environment was subject to considerable delay due to differences on the treatment of nuclear power between the two directorates (EC, 1989). In the transport sector, pressures for a paper on transport and cohesion were delayed by differences of opinion between the regional policy and transport directorates on how far regional development should be a concern in transport policy (EC, 1998e).

An example of differences in the College delaying affecting policy can be seen in the controversy over postal services liberalisation. Over the last three years the Commissioners for Internal Market and Competition have pushed for greater liberalisation while a significant core of other Commissioners – defined more by national interests – have been opposed. These disagreements came to a head in May 2000 when a much delayed proposal for liberalisation was held up by the continued opposition of the French and British Commissioners. The proposals eventually emerged but much scaled down from earlier drafts of proposals.

As to the Council, it is perhaps not surprising to note that proposals on energy and transport have usually ended up on the respective Council meeting agendas. Although energy and transport agendas have been marked by well-developed 'conventional wisdoms', in a number of areas, very large differences have existed between member states, thereby slowing down decision making.

Within the EP the key dynamics are those within the Committees responsible for a legislative proposal (where specific expertise and arguments will be to the fore) and in plenaries where votes are taken (where increasingly group preferences prevail). As noted before, however, the European Parliament's influence depends on a high degree of consensus amongst its members and this tends to dilute the influence of ideology in shaping Parliamentary positions (Bergman, 1997; Jacobs, Corbet and Shackleton, 1999).[10]

Ideas and Interests in EU Policy Making

As noted, the EU has become a focus for interest groups with very specific sectoral concerns to defend or promote. However, the issues which concern the EU policy-making process (as well as the style of policy making) require expert analysis and opinion: ideas and interpretations matter in the design of EU policies. In practice the distinction between interests and ideas is hard to maintain – one group's reasoned argument may be another's special pleading – and it is true that most groups seek to lend credibility to their arguments by drawing upon expert opinion. Moreover, in very specialised areas of policy, the 'stakeholders' may be the key sources of information and expertise. Nonetheless there are many instances where interventions by 'epistemic communities' (Richardson, 1996; Haas, 1992) seem to have played a role in shifting the direction of policy without there being an obvious economic interest at stake. Here we note the distinct roles of interest groups and other sources of knowledge and expertise.

It is a mark of the growing importance of decisions reached in the EU that so many interest groups now organise themselves to lobby at this level. Moreover, these interest groups have sought to take advantage of the multiple points of access within the EU system. This is not a passive process, however; indeed, interest groups have been encouraged to participate in the process by the European Commission (Schmitter and Streeck, 1991; Haas, 1963). From the earliest days of the EU, the Commission has fostered the development of European interest associations (often including them in committees formulating new proposals, in a few cases recognising their prior claims to set rules amongst themselves and even subsidising some organisations) as a way of enhancing the credibility of EU decision making (Greenwood, Rote and Gronit, 1998). Equally interest groups have been prepared to adapt themselves to this task, though the strength of many European associations is compromised by the concerns of many national associations to retain control over policy (with some associations being more concerned with intelligence gathering than opinion formation) and unwilling to provide substantial resources. Moreover in some cases, it is difficult for organisations to cohere due to divergent interests. In such cases actors may seek to lobby individually and to try to influence member states or form ad hoc coalitions with other interested parties (Pijnenburg, 1998; Coen, 1997).

Indeed the multiple paths for interest representation are a defining characteristic of the EU policy process (see Figure 2). An actor can choose to lobby singly or in a national or a European association and can target the Commission (or Commissioner), the Parliament or the national government. However, it is unclear whether these multiple points of access are an opportunity or a constraint in lobbying. With power diffused across different institutions, effective lobbying cannot be guaranteed. For their part the lobbied, especially the Commission, may be able to resist lobbying efforts and even exploit the variegated nature of the process to enhance their own preferences (Grande, 1996).

Figure 2.2   Multiple Targets and Modes of Interest Re-presentation

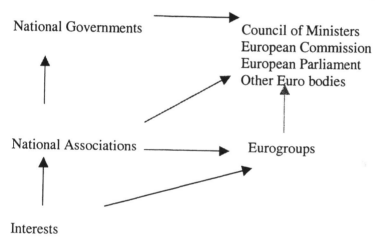

National Governments

Council of Ministers
European Commission
European Parliament
Other Euro bodies

National Associations ⟶ Eurogroups

Interests

Source: Adapted from Pijnenburg, 1998

In highly technical areas of policy-making the need to assess the likely impact of measures as well as the need for sectoral expertise has led to the creation of policy communities between commission officials, government officials, industry representatives and other specialists (with the roles being blurred by advisory relationships, consultancies and secondments). Information and expertise in interpreting that information are critical commodities in the EU policy process and increasingly groups that possess these commodities have been able to enhance their influence. So-called epistemic communities bring a range of broader, perhaps objective, values to the policy process and are able to make their opinions felt because of their very high credibility in their respective fields.

Box 4   Interests and Ideas in EU Policy Making

While the history of European energy and transport policy seems to indicate the strength of incumbent interests in both dominating the policy agenda and in blocking significant transfers of power to the EU level, developments in the recent past seem to suggest that the process is more open. In the application of environmental and market liberalisation policies non-incumbents have been successful in setting new agendas and in many cases have found willing and capable partners within the Commission to do so. Moreover as change takes place the ability of European associations to maintain coherence has been in some cases undermined by divergent interests. The main organisations in both electricity and air transport were split over market liberalisation with the UK members proposing more liberal measures than their continental colleagues (Grant, 1998; McGowan, 1998a). Moreover in both cases other associations representing other operators have emerged and gained influence. Indeed it seems that in some cases there seems to be an iterative process in some areas whereby the set pattern of lobbying as framed by initial market structure is changed by policies, which are then instrumental in bringing about further change within the market.

The role of ideas in the policy process can be seen most clearly in the proposals for Economic and Monetary Union. In this case the role of experts in framing the policy was absolutely critical to the development of the policy, just as their arguments were critical to the credibility of the claims made for a single currency. Moreover this widespread consensus extended further, particularly to questions such as the independence of central banks (Dyson and Featherstone, 2000).

A rather different example of the role of ideas in policy making is that of policy transfer. An example of this process – and its limits – is the idea of 'integrated resource planning' (IRP), a technique for appraising new investments in the energy sector which gave weight to the environmental benefits of otherwise more costly techniques of clean energy production or demand management vis-à-vis conventional supply side investments. In the 1990s environmental groups sought to persuade the Commission about the efficacy of this approach on the basis of its experience in the US. While they were successful in shaping a Commission legislative proposal, the established energy lobbies worked on the Commission and the member states to scale down the proposal, rendering it a relatively ineffectual decision (CEC, 1997). Arguably, a similar process is under way in the case of Commission proposals to introduce EU rules on congestion pricing in transport (CEC, 1995c).

That said, however, the Commission has been relatively open to other interests in devising policy (not least because of the different interests within the Commission itself). More generally, the experience of interest articulation within the EU seems to support the conventional wisdom concerning interest representation that the earlier lobbying takes place the more influential it is likely to be (Mazey and Richardson, 1993). Notwithstanding this general principle, powerful actors may be able to exercise vetos via clusters of national governments in extreme cases.

Multilevel Governance

The Commission's readiness to engage with a wide variety of interests extends to other tiers of government. The development of partnerships between the Commission and subnational governments have been seen as marking a qualitative shift in EU policy making towards a form of 'multilevel governance' (Marks *et al*, 1996). This approach does not reject the centrality of the state but it does challenge its monopoly, most notably in terms of the sharing of decision making both above and below the nation state. In terms of 'aboveness' it argues that supranational institutions do enjoy considerable independence and influence.

Box 5   The Importance of Multilevel Governance

How far has the multilevel policy making been apparent in EU energy and transport policies and how effective has it been? It is worth noting the diversity within these sectors in terms of modes, institutions etc. Thus some aspects are predominantly global (the oil sector) while others may be focused on local communities, whether with regard to a particular resource, or the delivery of services. A number of member states have maintained local utilities or transport services – indeed some aspects might be regarded as optimised in a local setting. The EU's role is potentially significant because of its role in financing regional development (with local infrastructures an important budgetary item) and in guaranteeing the provision of essential services (through the regulation of public service obligations). Moreover in both areas the Commission has been keen to facilitate the spread of 'best practice' in the ways that local agencies can encourage energy efficiency, alternative energy sources or public transport. To that end, the Commission has provided financial support for the development of regional planning agencies in both energy and transport (CEC, 1993 and 1998e). In each of these respects the Commission has engaged with local actors, governmental and non-governmental.

In terms of 'belowness' it sees subnational actors increasingly involved in decision making and their interests and tactics may contradict those of central government. Policy becomes a matter to be discussed and resolved on a variety of interconnected tiers (Wallace, 2000). While the model appears most appropriate for understanding those areas where interests, issues and resources converge and bring together the EU and local levels of government, it may also have a wider applicability, particularly with regard to the interaction between the EU and other Trans-national (or even global) regimes.

Europeanisation

The interplay between different levels of government have been a factor in concerning the extent of 'Europeanisation' of policies: how far and in what respects has the substance (and for some the style) of national policy making been Europeanised? By Europeanisation we understand a process whereby national policies converge around an agenda of priorities and instruments, which is set in a EU context (Ladrech, 1994, Laffan, O'Donnel and Smith, 2000). An important dimension of Europeanisation is of course the formal institutional setting, i.e., the various legislative initiatives that member states have to transpose along with the exercise of EU law as a constraint on the room for manoeuvre of national policy makers. However there is also a significant informal process at work, which we have already identified, i.e. the effect of repeated interactions amongst policy makers and those seeking to influence policy. Europeanisation also affects policy processes and styles; member states have had to adjust their respective administrative cultures and practices.[11] Europeanisation can be seen at work in such key policy areas as Economic and Monetary Union (Dyson and Featherstone, 2000) and in the provision of Cohesion and Structural Funds. A more curious example of Europeanisation has been in the case of competition policy where member states have steadily adapted their national legalisation to follow EU principles. Such has been the degree of Europeanisation that the Commission has proposed decentralising the administration of much of the policy, arguing that there is a shared culture amongst national authorities regarding the priorities of competition policy thereby obviating the need for control of policy at the centre (though it should be said that this view is not shared by those affected by competition policy).

Others are more sceptical of the Europeanisation thesis, pointing to the persistence of diversity in national policies and the failure of many states to follow EU policies even when they have apparently been agreed (see below). According to this view national approaches are resilient, reflecting both state traditions and real differences in interests in national settings. They particularly cast doubt on the grander accounts of Europeanisation, which touch on values and identity. Yet at more mundane levels, where the bulk of policy making takes place, there does appear to be some convergence amongst member states while in economic and corporate terms there does appear to be some Europeanisation of structures.

Implementation Gaps

An alternative interpretation holds that not only is Europeanisation failing to take place but that member states are seeking to defend their national interests by resisting the encroachment of EU policies into areas which are regarded as strategically or socially sensitive. Their efforts to do so are arguably assisted by the particular nature of the EU policy process. In a multilevel polity such as the EU the problem of ensuring effective implementation of agreed policies is likely to be greater than in centralised unitary systems. In the latter there is likely to be a greater degree of control and enforcement as well as fewer opportunities for opposition to manifest itself. By contrast a multitiered system provides both the opportunities to register opposition as well as a greater degree of distance in the policy process which permits shirking and/or reinterpretation (Richardson, 1996).

Failure to implement therefore may take the form of refusal to follow up on commitments (though this will eventually pose constitutional problems – how far can member states continue to belong to a club whose rules they persistently ignore?). More likely is the exercise of discretion in implementation to reinterpret or interpret narrowly the new policy or to signal preferences or support in ways that deter others from taking advantage of new rules. This is particularly so where the implementing authorities may have more or less vested interests in supporting particular operators within the sector (perhaps through public ownership).

Of course in areas of EU policy making it might be argued that there are means of redress and enforcement where implementation gaps persist. Those who stand to benefit from new policies may encounter the 'gap' and be prepared to challenge. The Commission's regulatory role is important in ensuring compliance (with legal action and Court sanctions as back-up). It may be therefore that at least some aspects of failure to implement are transitional hiccoughs, lags in the process of Europeanisation rather than challenges to it. That will in turn depend as much on the willingness of actors within the laggard states (NGOs, even independent agencies) to challenge any procrastination (Haas, 1998).

Box 6 Europeanisation ... or Not?

Just how much Europeanisation has taken place in the energy and transport sectors? After all, with long histories of attempts to devise common policies in both areas one might expect to see policies as comprehensively changed. In the transport sector, the drawn out process of market liberalisation has not yet transformed either the road haulage or air transport markets. However, while national structures persist there have been some signs of a gradual Europeanisation of market structures in the air transport sector and of the conduct of regulation. However, these changes may owe more to the globalisation of this sector in the former case and to the diffusion of EU competition policy in the latter (CAA, 1998).

Similarly in the electricity sector, the development of European rules has seen the establishment of European fora for regulators and the industry to find common European principles to operationalise the legislation. Yet other parts of the European electricity industry remain apparently impervious to change. This is even more the case for the European coal industry. Although ostensibly at the core of the European project (in the formation of the ECSC), the degree to which the European institutions have been able to assert control over national regimes for supporting the industry has been quite limited. Yet, while rules have been devised and work in some aspects of the industry (for example in managing the restructuring of areas where the industry has shut down) the core issue of subsidies has not been effectively addressed. Member states have to a large extent ignored the Commission (Newbury, 1995; Parker, 1997). The difficulties of ensuring effective implementation can be seen in the energy and transport cases. Arguably this has been a persistent problem in these areas reflecting the previously noted reluctance of member state governments to transfer sovereignty. In most cases however this reluctance has manifest itself earlier in the process. Failures to implement, it might be argued, are more likely to take place once the authority of the centre has been apparently accepted (or asserted); having failed to block policies, unwilling states may honour policy commitments in the breach rather than the observance. Thus in the field of air transport there were some concerns about the willingness of member states to first of all transpose legislation and then to follow the letter or spirit of the rules (as in the 'battle of Orly'). Similarly in electricity liberalisation, the main opponent of the EU policy, France continued to delay implementation until after the deadlines for transposition (Commission of the European Communities, 1998c). Elsewhere however, the realisation of policies has been assisted by the emergence of new entrants (and independent agencies) challenging the positions of incumbents and governmental supporters.

Explaining Policy Change

A central question for policy analysts is to explain the roots, orientation and process of change. We have already touched on some of the processes and actors involved in change but here we want to focus on three key issues. What are the formal and informal mechanisms which bring about change (building on the institutional discussions in the previous section)? What sorts of internal and external catalysts trigger change? How are the tensions between old and new policy agendas managed in practice? In the course of dealing with these three questions we also touch upon a series of other debates, notably those concerning the contribution of ideas and learning to policy change.

Formal and Informal Paths of Change

In our account of the evolution of the EU we touched upon the development of the acquis communautaire. This accumulation of legislation and judicial decisions stands as a register of the changes in EU priorities and processes. While this dynamic is incremental, there have been some important 'milestones' or quasi-constitutional elaborations of the acquis, notably the Intergovernmental Conferences. The latter have been best known as a venue for changing the balance of power between institutions but they have also been important for developing the scope of EU policy. To understand some of the politics of change in the EU we have to examine the dynamics of negotiations between governments in this forum as well as the extensive lobbying (where, in a sense, the EU institutions are themselves acting as lobbyists) which surrounds these events (Moravscic, 1999; Laursen and Hoonacker, 1995).

However we should not overstate the importance of these events. In some cases the actual effects of changes are not as dramatic as claimed; the shift from unanimity to QMV, for example, is better regarded as changing the incentive structures within which member states seek consensus rather than a radical shift to majoritarian decision making (Hix, 1999). Moreover, the process of policy change has not depended on these events to trigger policy. As noted, the Commission's entrepreneurialism has often manifest itself in creative interpretation of its existing powers (Heritier, 1999). Treaty provisions may take on very different meanings from those originally intended, while policies may themselves take on a momentum of their own, locking member states into a deepening of commitments which they may find hard to overturn. Indeed, the system of decision making means that there is a tendency towards irreversibility in the EU policy process (Pierson, 1996).

Box 7 Developing Policy – Formal and Informal Paths

The importance of informal mechanisms can be seen in the way in which the EU was able to develop an environmental policy without a treaty base: the policy began in the mid 1970s even though an environmental chapter was only included under the Single European Act. Similarly the policy of liberalisation, while the object of a difficult struggle between the Commission and its sympathisers in some member states on the one hand and a core of opposed member states on the other, effectively developed its own momentum: the initial phase of liberalisation created expectations of – and expanded the constituency for – further change.

The relative balance between formal and informal provisions has been apparent in energy and transport policy. The development of TransEuropean Networks as a strategic objective was reinforced by the inclusion of a provision in the Maastricht Treaty (even if it did not translate into substantial funds) and raised the profile in subsequent initiatives on transport funding (CEC, 1995b). A recognition of the importance of formal provisions was also apparent in the campaign to include a provision on public services, where establishing an article was deemed important to counter the 'liberalising' thrust of other treaty provisions, notably the Competition Rules (which were effectively impossible to amend given the support of many member states) (CEC, 1996c; Ross, 2000). Establishing a provision within the Treaty is desirable but may be neither necessary nor sufficient in the development of a competence. We have already noted the limited success of Common transport policy initiatives notwithstanding a clear commitment to such a policy within the treaty since the 1950s. Equally we have seen the development of numerous initiatives for an energy policy despite the lack of such a provision within the Treaty. To be sure the outcomes of such initiatives have often been limited but this seems to be due to more fundamental factors than the absence of a clear legislative base. Even so the Commission and some member states have continually pressed for the inclusion of a provision at each IGC and have been regularly rebuffed. By contrast the EU has developed a highly effective telecommunications policy without any sector-specific treaty provisions. Interestingly the original catalyst for this policy was the use of a very controversial treaty provision – article 86 (then 90) – which allows the Commission to impose policies directly on member states (a move which provoked hostility from some member states and a Court case which found in favour of the Commission). While the Commission proposed using this in the energy sectors, in the end it opted not to do so for fear of provoking further controversy even though it retained the threat to spur on liberalisation (McGowan, 1993).

Exogenous and Endogenous Sources of Change

Formal and informal paths of change, of course, do not happen in a vacuum. The changes can be responses to 'exogenous shocks' where new technologies or scientific discoveries, market changes or disruptions, or changes in the character or ideology of political leadership require policy-makers to alter the balance of policies and/or introduce new objectives. They can also be driven by endogenous processes where the policy makers themselves develop new ideas about policy priorities, perhaps in response to external changes, or recognise the limitations of existing policies. The boundary between these two categories is blurred, particularly as regards the impact of new ideas or interests.

These broader changes affected the policy communities within which EU policies were developed and the policy makers themselves. As already noted the development of the environmental debate brought in a range of NGOs and other groups seeking to shift the orientation of EU policies. However there was also a strong emphasis within these policy debates on establishing the credibility of new approaches. Arguably in such fields as market liberalisation and environmental protection distinct epistemic communities emerged to demonstrate the efficacy of new policies. A wider process of policy learning may also have been at work in terms of borrowing ideas and models from other sectors and countries and of recognising the limits of existing strategies and their incompatibility with prevailing conditions (Hood, 1994).

These factors may be symptomatic of an underlying dynamic or 'logic' to policy change within the EU. It may be that policy change reflects the prevalence of particular member state preferences and cultures: within the decision making bodies it may be that certain member states – or coalitions of states – will succeed in shaping policy (systemic logic). It may be that ideological shifts, reflected in the character of governments and of the European Parliament, will be strong enough to bring about policy change (partisan logic). It may be that policy changes arise as a result of changes within the sector in terms of the 'conventional wisdom' within that sector or because of particular problems encountered in that sector (sectoral logic).

The impact of these different logics is clear in the process of policy change within energy and transport policy. On the one hand political changes in the 1980s contributed to the rise of both environmentalism and neo-liberalism both in defining policy agendas and in determining particular decisions (for example the role of the European Parliament). The persistent opposition of some member states to liberalisation (in the case of France) and the differences in approach to environmental protection (North v South) suggest that policy on matters such as energy taxation owe more to systemic factors. In both cases however both the persistence of strong administrative cultures (in governments and the EU) and the transferability of new ideas from one country to another suggest a strong systemic logic at work.

Box 8　'Exogenous Shocks' in EU Policy Development

In the case of energy and transport, perhaps the most important catalysts for change have been in relation to the basic commodities upon which both sectors depend. Energy supply disruptions in the 1970s provoked an increased concern with security of supply. While this concern manifest itself in a range of policy responses (diplomacy and crisis management measures) it also gave an impetus to policies designed to improve the utilisation of energy. EU policy in this period focused on setting targets for member states and supporting programmes in energy efficiency and new technologies. These policies were harder to sustain after the 1986 collapse in energy prices but were revived as evidence for global warming highlighted the importance of energy efficiency in minimising emissions. This trend was reinforced by the rise of the Greens in much of the EU though arguably the more important political development was the spread of neo-liberal ideas and the emphasis on market making in both sectors. In this respect the changing configuration of member state governments was important in building coalitions within the Council of Ministers and providing policy exemplars on which EU policy proposals were based.

Managing Change: Conflicting Objectives, Missing Links and Common Policies

The impact of cross cutting policy agendas on specific areas of policy has a twofold effect: on the one hand it provides a momentum (and a rationale) for policy which had hitherto been undeveloped; on the other it may sit uneasily with some of the established policy objectives. There is in other words a risk of contradiction between different elements of the policy process. Thus the overall emphasis on market liberalisation (and particularly labour market flexibility) is in some instances at odds with the orientation of social policy in the Community. Similarly the macroeconomic rigour imposed by EMU constrains member states' capacity to follow certain policies without recreating the capacity (or its equivalent) at the EU level. There are problems for the Commission in reconciling conflicting objectives and addressing policy gaps. Piecemeal measures – the provision of exemptions in legislation – and overarching strategies – the development of Common policies – offer two ways of meeting this challenge though neither is entirely effective.

How can the Commission act to reconcile different objectives and fill in policy lacunae? In a sense the problem facing the Commission is a variant of the traditional negative-positive integration problem (Pinder, 1968). Negative integration, the removal of barriers to integration, is relatively easy to agree and enforce. Positive integration, the development of common policies is much more difficult, requiring more visible transfers of sovereignty from member states and often requiring substantial financial resources. In the current EU context we can see the problem in terms of the limits of the regulatory mode. It is relatively easy for the EU to agree to economic regulatory reform ('negative regulation') and possible to carry out social regulation ('positive regulation') though this is harder. This still leaves a substantial gap which is to some extent denied an EU solution because of the costs and commitments which are entailed. EU budgetary policies are confined to very specific areas of policy and the wider co-ordination of national policies (as opposed to their constraint) is very difficult to obtain.

Conclusions

So far this paper has dealt with some of the key interactions and some of the sources of policy change in the EU policy process. To conclude we consider how the EU fits into conventional stages of policy making (Hogwood and Gunn, 1984; Ham and Hill, 1993; Richardson, 1996). As noted earlier, such models may or may not be accurate descriptors of policy making in the real world. Yet they do highlight some important stages in the policy process. Borrowing from various conceptions in the policy analysis literature we focus on a number of key 'phases' in the process – formulation, consultation-negotiation, implementation-enforcement and evaluation – and consider how they are manifest in EU policy making.

Box 9 Reconciling Policy Conflicts and Filling Policy Gaps

How are conflicting objectives resolved and missing links addressed in EU transport and energy policy? There is clearly a problem in the case of energy policy where there has been a persistent tension between the traditional supply security agenda of energy policy with its emphasis on maintaining or developing Community energy resources which were not always the cheapest available, market liberalisation with its objective of lowest cost production and environmental protection which discouraged the use of scarce or polluting resources (CEC, 1988a, 1989, 1996a). In the transport field the 'infrastructure = growth' equation of regional transport development (CEC, 1998e) has been criticised by many environmentalists.

The traditional mode of reconciling policy differences – the Common Policy – has been to some extent discredited by the limited success of past attempts and by the perverse incentives and inefficiencies of the principal 'successful' common policy (the CAP). Such policies now tend to serve as general contextualisations or rationalisations for ongoing work programmes where the emphasis is now on a range of medium-range measures.

In the details of policy it has been possible to include derogations, exemptions or other special provisions within the overall legislation: in a sense this measure retains the regulatory mode but provides for other objectives to be allowed as 'special cases'. The electricity liberalisation directive includes provisions to protect environmentally friendly energy options (though these are criticised as inadequate). The application of state aids rules has tended to ignore government support for energy efficiency and cogeneration projects. However in many cases such measures are either temporary or counterproductive (by exposing the full costs of the measures involved, as in the case of nuclear power). Another approach has been to highlight particular objectives as taking primacy over others and requiring the 'integration' of those principles into all areas of EU policy. This is currently under way in the area of environmental protection (CEC, 1998b and 1998d) and is manifest in a recent initiative to encourage industry self-regulation and permit the automobile and energy industries to co-operate to reduce emissions of carbon dioxide from cars (CEC, 1998a). However, given the economic nature of European integration there are real limits to the extent of such integration. Attempts in the other direction, to 'marketise' externalities (placing a value on intangibles like supply security, congestion, emission damage etc and developing market mechanisms to operationalise these values) have encountered numerous technical problems and risk becoming compromises unacceptable to any parties.

Formulation of policy has traditionally been the primary responsibility of the Commission and one which it has pursued entrepreneurially and often with success. On a variety of major initiatives the Commission has been actively setting the agenda and reinterpreting its powers and competences to move policies forward and to establish an EU 'presence'. More recently however that role is less to the fore due to a combination of factors: the assertion by other institutions of policy competence (either instructing the Commission or even bypassing it in the case of 'intergovernmental' initiatives); the Commission itself becoming more involved in a monitoring or regulatory role, and the absence of sufficient resources to enable it to carry out all its tasks effectively. None of this is to suggest that the Commission is no longer a key partner in the policy process (although its credibility and morale have been affected by recent scandals) nor that accumulation of policy roles has come to an end, but it does mean a relative shift in the locus of formulation. Overall, however, the EU remains a dynamic venue for policy development, not least because of the possibilities for policy transfer between states and sectors.

Negotiation and Consultation involve an extensive interplay between the institutions, member states, interest groups and other social actors. The multilevel character of EU policy making is most apparent in this process with proposals oscillating between different levels and venues of government. The process primarily reflects the assertion of national and sectoral interests (the balance depending upon the nature of the policy) and less so ideological positions (though 'ideational' differences are important), a consequence of the technical, 'low politics' nature of many EU policies. However, as the scope of EU policies has widened and the consequences and potential of 'technical' measures has become apparent there has been a certain broadening-out of the constituencies involved in policy making. This trend can be seen in the increasing role of the European Parliament (which is arguably more active and independent than most national legislatures) as well as in the involvement of other levels of government and other types of social movement beyond economic interest groups. These developments and the European Commission's generally positive response to them can be seen as addressing criticisms that there is a democratic deficit in EU policy-making: although the European institutions are relatively open to proposals and ideas, many have criticised the lack of transparency in decision making, particularly within the Council of Ministers. In making the EU more accountable, however, there may be some cause for concern at the impact on policy effectiveness: will an 'overload' of demands upon the EU create new logjams?

Implementation and Enforcement issues arise due to the multilevel character of EU policy making and the resource constraint at the EU level. Indeed our focus shifts back to the member states given that the results of negotiation have to be implemented at the national (or subnational) level. As we have noted there may be problems of compliance and the supranational role becomes one of monitoring and regulation. It is in this realm that the European legal system – particularly the Court – is most important, providing a mechanism of enforcement vis-à-vis governments as well as firms. The role of interest groups and other social movements is less obvious in this realm, except insofar as they can bring complaints and expose shortcomings and non-compliance.

Evaluation in the European Union is both an extensively practiced and a relatively under-explored process. A variety of mechanisms ranging from more or less rigorous scrutiny and audit procedures which mainly focus upon financial matters to assessments of policy impact are conducted both by the Commission and by specialist European agencies such as the Court of Auditors. The purposes of such evaluations also vary – with some apparently designed to demonstrate the success of a policy (or to explain why the results are less positive than originally envisaged) while others (particularly from outside the Commission) to question the effectiveness or even desirability of the policy (though there has been a trend towards 'sunsetting' particular policy measures which are deemed to have outlived their usefulness). The process is not confined to the European institutions, with many non governmental organisations and pressure groups producing their own analyses of policy impacts. They do so – as do many of those engaged in evaluations – as the first step in the next stage of policy making, closing the loop of the EU policy process.

Notes

1 The term implies as much the sharing of sovereignty as its transfer, a positive-sum rather than a zero-sum game.

2 An additional forum for consultation was created in the form of the Economic and Social Committee, a body representing different sectional interests though with a focus on business and labour. Although modelled on 'corporatist'-type bodies in many European countries, this body was to prove less significant than was intended.

3 Similarly, the analysis of the EU has been largely informed by approaches which tend to emphasise one or the other of these as the primary dynamic, with one or other of these 'schools' prevailing according to these events, while more recently there has been a shift towards approaches which concentrate on the functioning of the institutions rather than the locus of power. See Hix, 1995; Wallace and Wallace, 2000 and Rosamond, 2000.

4 In practice, however, policies in these areas had existed beforehand, a reflection perhaps of the Commission's 'entrepreneurship'.

5   Mainly due to their economic similarities to the core of the EU and the 'transition' role of the European Economic Area, an agreement signed between the European Community and most of the members of the European Free Trade Area (Begg, 1992).

6   Though it must be said that these reforms are facing considerable opposition from the Commission workforce. See the weekly newspaper *European Voice* for details.

7   These issues will be up for debate at yet another IGC to be convened in 1004. On the Nice Summit see European Council (2000).

8   Although European integration has served as a model for regional cooperation elsewhere (Nye, 1969; Mattli, 1999).

9   This idea has been a more or less explicit implication of various proposals for a 'Europe of the Regions'. For a critical view see Jeffrey (2000).

10   Though some argue that a 'left-right' division in EP votes is becoming more apparent (Hix, 1999).

11   Arguably this is best seen in the case of new member states.

References

Abbati, C. (1987), *Transport and European Integration*, Commission of the European Communities, Brussels.

Allen, D. (2000), 'Cohesion Policy' in Wallace and Wallace (eds), *Policy Making in the European Union*, Oxford University Press, Oxford.

Alter, K. (1998), 'Who are the 'Masters of the Treaty'? European Governments and the European Court of Justice', *International Organization*, Vol. 52, No. 1.

Alting von Geusau, F. (1975), 'In Search of a Policy' in Adelman, M. and Alting von Geusau, F. (eds), *Energy in the European Communities*, Sijthoff, Leyden.

Argyris, N. (1989), 'The EEC Rules of Competition and the Air Transport Sector', *Common Market Law Review*, Vol. 26, No. 1.

Begg, D. *et al.* (1992), 'Is Bigger Better? The Economics of EC Enlargement', *Monitoring European Integration*, Vol. 3, CEPR, London.

Bergman, T. (1997), 'National Parliaments and EU Affairs Committees: Notes on Empirical Variation and Competing Explanations', *Journal of European Public Policy*, Vol. 4, No. 3.

Burley, A. and Mattli, W. (1993), 'Europe Before the Court', *International Organization*, Vol. 47, No. 1.

Cini, M. (1997), *The European Commission: Leadership, Organisation and Culture in the EU Administration*, Manchester University Press, Manchester.

Civil Aviation Authority (1998), *The Single European Aviation Market – the First Five Years*, Cheltenham: Westwood.

CEC (1985), *Completing the Internal Market*, Commission of the European Communities, Brussels.

CEC (1988), *The Internal Energy Market*, Commission of the European Communities, Brussels.

CEC (1988), *Review of Member States' Energy Policies*, Commission of the European Communities, Brussels.

CEC (1989), *Energy and the Environment*, Commission of the European Communities, Brussels.

CEC (1992), *The Future Development of the Common Transport Policy*, Commission of the European Communities, Brussels.

CEC (1993a), *Energy and Economic and Social Cohesion*, Commission of the European Communities, Brussels.

CEC (1994), *For a European Union Energy Policy – Green Paper*, Commission of the European Communities, Brussels.

CEC (1995a), *The Development of the Community's Railways*, Commission of the European Communities, Brussels.

CEC (1995b), *The Common Transport Policy Action Programme 1995-2000*, Commission of the European Communities, Brussels.

CEC (1995c), *Towards Fair and Efficient Pricing in Transport*, Commission of the European Communities, Brussels.

CEC (1996a), *White Paper on Energy Policy*, Commission of the European Communities, Brussels.

CEC (1996b), *Revitalising Europe's Railways*, Commission of the European Communities, Brussels.

CEC (1996c), *Services of General Economic Interest in Europe*, Commission of the European Communities, Brussels.

CEC (1996d), *Proposal for a Council Decision Concerning the Organisation of Cooperation Around Agreed Community Energy Objectives*, Commission of the European Communities, Brussels.

CEC (1997), *An Overall View of Energy Policy and Actions*, Commission of the European Communities, Brussels.

CEC (1998a), *Implementing the Community Strategy to Reduce CO2 Emissions From Cars. An Environmental Agreement With the European Automobile Industry*, Commission of the European Communities, Brussels.

CEC (1998b), *Partnership for Integration: A Strategy for Integrating Environment into EU Policies*, Commission of the European Communities, Brussels.

CEC (1998c), *State of Liberalisation of Energy Markets*, Commission of the European Communities, Brussels.

CEC (1998d), *Strengthening Environmental Integration Within Community Energy Policy*, Commission of the European Communities, Brussels.

CEC (1998e), *Transport and Cohesion*, Commission of the European Communities, Brussels.

CEC (2000a), *Annual Report on Competition Policy*, Commission of the European Communities, Brussels.

CEC (2000b), *Reforming the Commission*, Commission of the European Communities, Brussels.

Coen, D. (1997), 'The Evolution of the Large Firm as a Political Actor in the European Union', *Journal of European Public Policy*, Vol. 4, No. 1.

Coombes, D. (1970), *Politics and Bureaucracy in the European Community, A Portrait of the Commission of the E.E.C.*, Allen and Unwin, London.

Diebold, W. (1959), *The Schuman Plan: A Study in Economic Cooperation, 1950-1959*, Praeger, Oxford.

Dinan, D. (1999), *Ever Closer Union*, Macmillan, Basingstoke.

Dyson, K. and Featherstone, K. (2000), *The Road to Maastricht*, Oxford University Press, Oxford.

Earnshaw, D. and Judge, D. (1995), 'Early Days: The European Parliament, Co-decision and the European Union Legislative Process Post-Maastricht', *Journal of European Public Policy*, Vol. 2, No. 4.

European Council (2000), *Presidency Conclusions*, European Council, Nice.

Garrett, G. (1995), 'The Politics of Legal Integration in the European Union', *International Organization*, 1995, Vol. 49, No. 1.

Gatsios, C. and Seabright, P. (1989), 'Regulation in the European Community', *Oxford Review of Economic Policy*, Vol. 5, No. 2.

George, S. (1996), *Politics and Policy in the European Community*, University Press, Oxford.

Gillingham, J. R. (1991), *Coal Steel and the Rebirth of Europe*, Cambridge University Press, Cambridge.

Grande, E. (1996), 'The State and Interest Groups in a Framework of Multi-Level Decision-Making: The Case of the European Union', *Journal of European Public Policy*, Vol. 3, No. 3.

Grant, W. (1987), *Business and Politics in Britain*, Macmillan, Basingstoke.

Greenwood, J., Grote and Ronit (eds) (1998), *Organised Interests and the European Community*, Macmillan, Basingstoke.

Gwilliam, K.M. and Mackie, P.J. (1975), *Economics and Transport Policy*, Allen & Unwin, London.

Haas, E.B. (1963), *The Uniting of Europe: Political, Social, and Economic Forces, 1950-1957*, University Press Stanford, Stanford.

Haas, P. (1992), 'Epistemic Communities', *International Organisation*, Vol. 46, No. 1.

Haas, P. (1998), 'Compliance With EU Directives: Insights From International Relations and Comparative Politics', *Journal of European Public Policy*, Vol. 5, No. 1.

Ham, C. and Hill, M. (1993), *The Policy Process in the Capitalist State*, Harvester, Brighton.

Hayes Renshaw, F. and Wallace, H. (1996), *The Council of Ministers of the European Union*, Macmillan, Basingstoke.

Heritier, A. (1996), 'The Accommodation of Diversity in European Policy-Making: Regulatory Policy as Patchwork', *Journal of European Public Policy*, Vol. 3, No. 3.

Heritier, A. (1999), *Policy Making and Diversity in Europe*, Cambridge University Press, Cambridge.

Hix, S. (1995), 'Approaches to the Study of the European Community', *West European Politics*, Vol. 17, No. 1.

Hix, S. (1999), *The Political System of the European Union*, Macmillan, Basingstoke.

Hoffmann, S. (1966), 'Obstinate or Obsolete? The Fate of the Nation-State and the Case of Western Europe', *Daedalus*, No. 95.

Hogwood, B. and Gunn, L. (1984), *Policy Analysis for the Real World*, University Press Oxford, Oxford.

Hood, A. (1994), *Explaining Policy Reversals*, Oxford University Press, Oxford.

Jacobs, F., Corbett, R. and Shackleton, M. (2000), *The European Parliament*, Harper, London.

Jeffery, C. (2000), 'Sub-National Mobilization and European Integration: Does it Make any Difference', *Journal of Common Market Studies*, Vol. 38, No. 1.

Junge, K. (1999), *Flexibility, Enhanced Co-operation and the Treaty of Amsterdam*, Kogan-Page, London.

Keohane, R. and Hoffman, S. (1992), 'Institutional Change in Europe in the 1980s', in Keohane, R.O. and Hoffman, S. (eds), *The New European Community: Decisionmaking and Institutional Change*, Westview, Boulder.

Ladrech, R (1994), 'Europeanisation of Domestic Politics and Institutions: the Case of France', *Journal of Common Market Studies*, Vol. 32, No. 1.

Laffan, B. (1997), 'From Policy Entrepreneur to Policy Manager: The Challenge Facing the European Commission', *Journal of European Public Policy*, Vol. 4, No. 3.

Laffan, B. (1998), 'The European Union: A Distinctive Model of Internationalization', *Journal of European Public Policy*, Vol. 5, No. 2.

Laffan, B., O'Donnel, R. and Smith, M. (2000), *Europe's Experimental Union*, Routledge, London.

Laursen, F. and Vanhoonaker, S. (1992), *The Intergovernmental Conference on Political Union*, EIPA, Maastricht.

Lewis, J. (1998), 'Is the "Hard Bargaining" Image of the Council Misleading? The Committee of Permanent Representatives and the Local Elections Directive', *Journal of Common Market Studies*, Vol. 36, No. 4.

Lindberg, L. and Scheingold, S. (1970), *Europe's Would-be Polity: Patterns of Change in the European Community*, Prentice-Hall, Hemel Hempstead.

Machlup, F. (1977), *A History of Thought on European Integration*, Macmillan, Basingstoke.

Majone, G. (1994), 'The Rise of the Regulatory State in Europe', *West European Politics*, Vol. 17, No. 3.

Mancini (1989), 'The Making of a Constitution for Europe', *Common Market Law Review*, Vol. 26, No. 4.

Marks, G., Hooghe, L. and Blank, K. (1996), 'European Integration from the 1980s: State-Centric v Multi-level Governance', *Journal of Common Market Studies*, Vol. 34, No. 3.

Mattli, W. (1999), *The Logic of Regional Integration*, Cambridge University Press, Cambridge.

Mazey, S. and Richardson, J. J. (eds) (1993), *Lobbying in the European Community*, Oxford University Press, Oxford.

McGowan, F. (1990), 'Conflicting Objectives in EC Energy Policy', *Political Quarterly Special Edition – The Politics of 1992*.

McGowan, F. (1993), *The Struggle for Power in Europe 1993*, RIIA, London.

McGowan, F. (1998a), 'Transport Policy' in El-Agraa, A. (ed.), *The European Union: History, Institutions, Economics and Policies*, Prentice Hall, Hemel Hempstead.

McGowan, F. (1998b), 'Energy Policy' in El-Agraa, A. (ed.), *The European Union: History, Institutions, Economics and Policies*, Prentice Hall, Hemel Hempstead.

McGowan, F. (1999), 'Globalisation, Regional Integration and the State' in Shaw, M. (ed.), *Globalisation and Politics*, Routledge, London.

McGowan, F. and Seabright, P. (1995), 'Regulation in the European Community and its Impact on the UK' in Bishop, M., Kay, J. and Mayer, C. (eds), *The Regulatory Challenge*, Clarendon, Oxford.

McGowan, F. and Wallace, H. (1996), 'Towards a European Regulatory State', *Journal of European Public Policy*, Vol. 3, No. 4.

Metcalfe, L. (1996), 'The European Commission as a Network Organisation', *Publius*, Vol. 26, No. 3.

Moravcsik, A. (1991), 'Negotiating the Single European Act', *International Organisation*, Vol. 45, No. 1.

Moravcsik, A. (1999), *The Choice for Europe*, Cornell University Press, Cornell.

Neunreither, K. (1994), 'The Democratic Deficit of the European Union', *Government and Opposition*, Vol. 29, No. 3.

Newbery, D. (1995), 'Removing Coal Subsidies', *Energy Policy*, Vol. 23, No. 6.

Nye, J. (ed.) (1968), *International Regionalism*, Little Brown, Boston.

O'Leary, S. (1995), 'The Relationship Between Community Citizenship and the Protection of Fundamental Rights in Community Law', *Common Market Law Review*, Vol. 32, No. 2.

Parker, M. (1997), *The Politics of Coal's Decline*, RIIA, London.

Pelkmans, J. (1984), *Market Integration in the European Community*, Sijthoff, Leyden.

Peters, B.G. (1992), 'Bureaucratic Politics and the Institutions of the EC', in Sbragia, A. (ed.), *Europolitics*, Brookings, Washington.

Peterson, J. (1995), 'Decision-Making in the European Union: Towards a Framework of Analysis', *Journal of European Public Policy*, Vol. 2, No. 1.

Pierson, P. (1996), 'The Path to the European Integration – A Historical Institutionalist Analysis', *Comparative Political Studies*, 1996, Vol. 29, No. 2.

Pijnenburg, B. (1998), 'EU Lobbying by ad hoc Coalitions: An Exploratory Case Study', *Journal of European Public Policy*, Vol. 5, No. 2.

Pinder, J. (1968), 'Positive Integration and Negative Integration: Some Problems of Economic Union in the EC', *World Today*, Vol. 24, No. 3.

Radaelli, C. (2000), *Technocracy in the European Union*, Longman.

Richardson, J. (ed.) (1996), *European Union. Power and Policy-Making*, Routlegde, London.

Rosamond, B. (2000), *Theories of European Integration*, Macmillan, Basingstoke.

Ross, G. (1995), *Jacques Delors and European Integration*, Polity Press, Cambridge.

Ross, M. (2000), 'Article 16 EC and Services of General Interest: From Derogation to Obligation', *European Law Review*, Vol. 25, No. 1.

Sandholtz, W. and Zysman, J. (1989), '1992 – Recasting the European Bargain', *World Politics*, Vol. 42, No. 1.

Scharpf, F. (1988), 'The Joint Decision Trap: Lessons from German Federalism and European Integration', *Public Administration*, Vol. 66, No. 3.

Schmitter, P. and Streeck, W. (1991), 'From National Corporatism to Transnational Pluralism', *Politics and Society*, Vol. 19, No. 2.

Spierenburg, D. and Poidevin, R. (1994), *The History of the High Authority of the European Coal and Steel Community: Supranationality in Operation*, Weidenfeld and Nicolson, London.

Spinelli, A. (1966), *The Eurocrats*, John Hopkins, Baltimore.

Tsebelis, G. (1994), 'The Power of the European Parliament as a Conditional Agenda Setter', *American Political Science Review*, Vol. 88, No. 1.

Tsoukalis, L. (1997), *The New European Economy Revisited*, Oxford University Press, Oxford.

Wallace, H. and Wallace, W. (eds) (2000), *Policy-making in the European Union*, Oxford University Press, Oxford.

Weiler, J. (1994), 'A Quiet Revolution – the European Court of Justice and its Interlocutors', *Comparative Political Studies*, Vol. 26, No. 4.

Wessels, W. (1997), 'The Growth and Diffusion of Multilevel Networks', in Wallace, H. and Young, A.R. (eds), *Participation and Policy-Making in the European Union*, Clarendon Press, Oxford.

Westlake, M. (1994), *A Modern Guide to the European Parliament*, Pinter, London.
Wyatt, D. and Dashwood (1987), *European Community Law*, Sweet and Maxwell, London.

3 The EU Enlargement and its Impact on European Policies

GERHARD RAMBOW

The process of accession of ten Central and Eastern European States, Cyprus and Malta and, finally, Turkey presents the European Union with a rare but also promising challenge.

None of the earlier accessions of new Member States to the European Communities/European Union was quite comparable to the present accession process:

- The number and size of the countries concerned, if taken together, exceeds that of earlier accessions. Without Turkey, their population is almost 30 per cent of the population of the present EU Member States (i.e. 104 million as against 374 million); including Turkey the number is close to 45 per cent.
- What might be of greater importance: The prospective new Member States from Central and Eastern Europe are still in a difficult process of restructuring their industry and agriculture and of adapting them to the market economy, taking into account the Common Agricultural Policy.
- The economic development and prosperity of the CEECs are far behind that of the EU-average, and even below that of the least developed actual Member States. All of the prospective Member States have a GDP/capita well below 75 per cent of the present EU average. Only the Czech Republic and Slovenia display a GDP/capita of a similar order to that of Greece and Portugal.
- In addition, it must be kept in mind that the more fully developed the EU integration, the more difficult accessions will be. It is obvious that accepting the acquis presents countries with greater problems if this acquis consists of ever more rules and encompasses more thoroughly elaborated policies.

The latter was not the case for Austria, Finland, and Sweden as they were already Members of the European Economic Area and included into the EC's internal market (with the exception of agriculture).

Measured exclusively with regard to change in population size, the accession of Denmark, Ireland, and the United Kingdom represented a comparable step: these three countries had together a population of roughly 66 million which was a third of the then population of the six Members of the EEC. This earlier accession did not present any of the problems of substance that the EU is facing now, however; at that time – the year 1973 – the EEC was much less developed, and the economic development of the 'new' Member States was, with the exception of Ireland, comparable to that of the 'old' Member States.

Accession into a highly-developed EU puts heavy demands on the applicant states. Besides having to harmonize great parts of their legal order with that of the EU, they are expected to adjust their policies in all areas of relevance to the EU, and to prepare their economies to be able to withstand competition in the Single Market. A great part of this process in the applicant states is inevitable and part of their 'natural' development following the breakdown of the COMECON order.[1] Still, the time constraint and the obligation to adopt to the EU's legislation and policies – whilst not yet being a partner – presents political and practical problems to these countries. This is not always fully appreciated within the EU.

The EU supports the preparation of these countries for accession by way of a substantial programme. The Essen European Council of 1994 agreed on an Accession Strategy. The Accession Partnerships concluded with each country set the concrete framework for this assistance. The EU financial framework for the period 2000-2006 (Berlin European Council) provides a total amount of € 21.840 million for pre-accession aid.

This article deals with the impact of the Eastern enlargement on the EU and its policies. Among the policy areas it concentrates in particular on the budget as the overall framework, on the cohesion policy, including structural measures, and on agricultural policy. It also discusses, albeit more briefly, issues of competition, environmental policy, social policy and transport policy.[2]

I begin with some general observations on the general issues of diversity, policy implementation, subsidiarity and flexibility and briefly discuss the nature of the Acquis which sets the framework for accession.

The sheer size of the future Union will make it even more necessary than it is at present to think about the proper way for the EU to function in an optimal fashion. This will, or should, also influence EU policies and their application. The discussion of a 'core-Europe' or a two-tier-Union is to be seen in this context.

It is not only the size of the 'new' Union but its much greater diversity – especially if one includes Turkey – that is at issue here. This diversity cannot be measured exclusively in economic terms, or by the relative economic development and/or prosperity of the countries or regions concerned; it clearly also has a 'cultural' dimension. This greater diversity will require adjustments in the way the Union is governed.

As far as economic criteria are concerned, one may safely assume that the now considerable differences in economic development, wage levels, or infrastructure will decrease over time. But it would be wrong to expect them to disappear altogether. Differences exist within the present Union as well. The development of the EU policies will have to take this into account.

In order to assess these problems correctly, it is also important to recall that the task of developing agrarian societies or regions is fundamentally different from that of re-structuring industrial societies or regions. The integration of the former GDR territories into the EU made – and still makes – this quite clear. Policy on the national and on the EU level did not always take this into account sufficiently.

EU Policy and Diversity

The greater diversity of the EU following enlargement will influence policy decisions and their content. This can be done in a number of ways and will undoubtedly differ from case to case:

- the EU may choose to incorporate elements of diversity in its actions;
- alternatively, the EU might limit itself to framework decisions and leave it to the Member States to apply more concrete rules;
- or else the EU could set only minimum standards, leaving considerable latitude to Member States.

Such practices are not new to the EU. They have become more common recently.

The Position Papers of the applicant states provide numerous examples where greater diversity in EU rules and policies might be necessary or advisable. They also suggest that this is a much broader and more general challenge with which the EU will have to cope in the future.

One relevant issue is the cost of policy measures and their prioritisation. Obviously the prioritisation of policy objectives or policy measures looks different in developed and prosperous economies than it does in less developed and less prosperous economies, including even some of the present Member States. This aspect is of importance for EU measures that require costly investments for their application, such as environmental protection. The same consideration applies to rules of a more general nature leading to economic costs for enterprises, such as rules on social protection. This has to be taken into account when the EU sets its priorities and decides on its policies.

The EC Treaty refers to considerations of this kind in several of its provisions:

- Art. 137 para 2 stipulates that in the areas of social policy governed by qualified majority decisions should have 'regard to the conditions and technical rules obtaining in each of the Member States', and that directives 'shall avoid imposing administrative, financial and legal constraints in a way which would hold back the creation and development of small and medium-sized undertakings'.
- Art. 175 para 5 EC Treaty (environmental measures) allows for temporary derogations and/or financial support from the Cohesion Fund in cases where the measure 'involves costs deemed disproportionate'.

The principle of proportionality (see Art. 5 para 3 EC Treaty) is by itself not a sufficient rule for the differentiated assessment necessary in these cases. Yet there is no necessity to establish a new legal base for taking diversity into account, as the Community Institutions (Parliament, Council, Commission) have the necessary power of discretion in decision-making.

The need for differentiation, however, is likely to be strengthened with Eastern enlargement. The various areas of EU policy and law lend themselves to such differentiations in different degrees. In some cases, this will present the EU with a difficult choice. Diversity must not endanger the common framework.

It will be of central importance to the EU that the smooth functioning of the internal market without frontiers be preserved.[3] Taking diversity into account in other areas than the internal market will be less problematic. Accession could lead to a more general policy in the EU to avoid uniform and detailed rules where they are not necessary.

Nevertheless we need to be aware of the fact that taking diversity into account might run into opposition within the EU. Opposition might come from Member States fearing competitive disadvantages as a result of the diversity in regulations. Or it might come from Members States treated differentially to their disadvantage, for instance if the amount of direct payments to farmers were made to correspond to the average income or standard of living within a country.

There are several policy areas where such a differentiation would not endanger integration, as in areas of environmental and social policy not directly linked to competition and the internal market or in the structural policies.

If the principle of diversity were applied more generally by the EU as a consequence of enlargement this would have a wholesome effect on EU legislation. In assessing this development one should keep in mind:

- Integration does not equal uniformity, and the degree of integration cannot be measured by the degree of uniformity.
- It is the EU Institutions themselves which decide on this differentiation and over which they keep 'control'.
- Uniformity in areas where it is not justified and where it leads to unacceptable costs (in a wider sense) would be harmful for the further process of integration.

Policy Implementation

Greater differentiation appears inevitable also with regard to policy implementation. Different traditions, especially with regard to the concept of law and legality, are an important factor within the Union already today. This is, however, not always talked about openly, nor are the necessary conclusions drawn from it for the formulation of EU policy.

In certain areas a homogeneous application of EU law is more important, if not essential, for the functioning of the EU than in others.

There are important areas of EU policy where the application rests with EU institutions themselves. Leaving the European Central Bank aside (because it is limited to the Members of EMU), one may cite the rules on competition, including the control of state aid, as well as the execution of certain financial programmes, such as research and development.

In general however, the application of EU law, or of national law based on EU legislation, or of EU policies is entrusted to national administrations. This is also the case, for instance, of the Common Customs Tariff, the CAP, or the Structural Funds. Modalities of co-operation with the EU Commission are established, albeit only for certain areas and to a differing degree.

It is illusionary to believe that ever tighter controls will ensure a homogeneous application of EU legislation in all 'corners' of the EU. Even George Orwell's 'Big Brother' would not be able to guarantee this. The concept of the EU itself, its laws and its policies must take account of such – likely – 'imperfections' in their execution.

There are many rules in the CAP but also in other areas of EU law which are so complicated that their uniform application is very unlikely even within the present Union. It is to be welcomed that a simplification of EU legislation has become an EU policy. The application of this principle needs to be pushed much further.

Subsidiarity

The greater Union will have to take the application of subsidiarity much more seriously. The opinions on its importance and relevance differ widely as is reflected, among others, in the Protocol No. 30 on the Application of the Principles of Subsidiarity and Proportionality.

According to the principle of subsidiarity, as inscribed in the EC-Treaty, the Community

> shall take action…only and insofar as the objectives of the proposed action cannot be sufficiently achieved by the Member States and can therefore, by reason of the scale or effects of the proposed action, be better achieved by the Community (Art. 5 para 2 EC-Treaty).

A much greater and more diverse Union will have to consider with greater circumspection than it does at present whether it is really the case that a proposed action can 'be better achieved by the Community'. This element of the subsidiarity clause may gain a new significance in the greater Union. One will have to weigh the 'sufficiently achieved' against the 'better achieved' and decide what is preferable: national actions with some imperfections or an EU action beset by – unavoidable – shortcomings? Apart from the 'yes-or-no' of an EU action, subsidiarity also influences the *kind* of action the Union may take in a given area (for instance, framework decisions, minimum standards).

Flexibility

The Treaty of the European Union and the EC Treaty contain provisions on 'closer co-operation'. These provisions allow a group of Member States to establish among themselves a closer co-operation on certain issues within the framework of the competences of the EU. Such closer co-operation is currently possible only as a last resort 'where the objectives of the (...) Treaties could not be attained by applying the relevant procedures laid down therein' (Art. 43 EU Treaty). The Treaty of Nice will make closer cooperation easier to achieve: closer cooperation may be engaged 'when it has been established within the Council that the objectives of such cooperation cannot be attained *within a reasonable period* by applying the relevant provisions of the Treaty' (Treaty of Nice, provisional text of 12 December 2000, Enhanced Cooperation, General Principles, Clause B – my italics).

The greater diversity after the accession of the applicant States is no reason to see in this so-called 'flexibility-clause' an appropriate path for the further development of EU policies. Quite to the contrary, if these provisions were used for this purpose this would be dangerous for the Union's cohesiveness.

The task that the EU faces after accession is to take diversity more strongly into account when formulating EU rules applicable throughout the Union. The problems connected with that cannot be solved by using the flexibility clause, or by instituting a two-tier-Community. What should be avoided under all circumstances is to treat the new Member States *en bloc* differently from the old Member States. Taking diversity seriously into account will cut across this line.

The recently renewed discussions on forming a core-group within the EU should be of concern to all those in Europe who think of the EU as one multilateral Union where each Member State is an equal member, i.e. a member on the same terms as all others. This multilateral Community faces a real danger if a certain number of its members form a 'centre of gravity' and agree to 'speak with one voice (...) on as many issues as possible' as proposed by J. Fischer, the German Foreign Minister, in a speech given at

Humboldt University on the 12th May, 2000. This would lead to a voting block which is likely to dominate the EU. I believe this to be incompatible with the spirit of the Union.

It seems to be the case that interested circles within the EU take accession as a pretext for revitalizing such ideas. In that, they present Europe with a fundamental choice.

Accession need not have this type of impact on the EU and its structure. The new Member States do not want to be left in a second tier of the Union, even if they were to find themselves in that tier in the company of others.

The Acquis and the Development of EU Policies

As a general rule, the new Member States will have to accept the acquis of the EU as it stands at the time of accession.

The *Acquis Communautaire* seems to take on almost magical qualities at times. Art. 2 of the EU Treaty counts among the objectives of the Union 'to maintain in full the acquis communautaire and build on it'. The term 'acquis' does not have the same meaning in all the contexts in which it is used, however.

Art. 2 of the EU Treaty can only refer to important rules but not to every regulation, directive, or decision ever adopted by the Union. The provision itself refers to necessary revisions of the acquis. Such revisions might be quite substantial, as is exemplified by the ongoing reform of the Common Agricultural Policy.[4]

In the context of accession, the acquis has a very general meaning: it includes all EU legislation and policies already established as well as all the revisions which might take place before the day of accession. This approach of the EU toward the accession negotiations precludes that the new Members can request fundamental changes of the acquis *during* the negotiations.

The EU might find it necessary to adapt the acquis on its own accord when new Members join. One may assume that such changes prior to accession would be done to safeguard the interests of the present EU Member States and that they might not be so much in the interest of the new Members.

One case in point is the (new) rule established by the Berlin European Council according to which the 'total receipts in any Member State from structural operations (...) should not exceed 4 per cent of national GDP' (Berlin European Council, 24 and 25 March 1999, Presidency Conclusions, Doc. SN 100/99, No. 46). This threshold was chosen carefully so as not to limit the actual receipts of the present Member States but to put an effective ceiling on receipts for new Member States with their – as yet – lower GDP/capita. The Common Agricultural Policy presents special problems; these are discussed in a separate section of this chapter.

Changes or adaptations might also be a consequence of accession and be decided upon after new Members have acceded to the Union. The new Members will then have a voice in the decision making bodies and their interests will have to be taken into account to the same extent as the interest of the 'old' Member States.

Former accessions provide some examples: Thus, after the accession of Spain and Portugal the CAP was more strongly developed with regard to the so-called 'southern' products (i.e. fruit, vegetables, wine, olive oil, tobacco). The southern Member States had acquired a blocking minority which enabled them to press for the full extension of the CAP to southern products. This was so successful that by now the Member States in the South (Greece, Italy, Portugal and Spain) receive about one third of the transfers under the CAP while contributing only about 22 per cent to the EU budget (Report of the Court of Auditors, 1998).

The establishment of Objective-6-regions under the Structural Funds when Austria, Finland, and Sweden joined the EU may be cited as another example even though this was part of the Accession Treaty. The Berlin European Council decided in 1999 to include these areas under Objective 1 of the Structural Funds Regulation.

Budget

It is useful to start the discussion of policies affected by accession with a look at the budget. Apart from the more structural questions of how to run the greater Union, it is most likely that the budget and policies depending on expenditures are most strongly affected by accession. It is with good reason, I believe, that the future development of Cohesion Policy and the Common Agricultural Policy might feel the impact of accession more than other EU policies.

Present Financial Framework

The Berlin European Council approved the Financial Perspective for the years 2000 through 2006 which includes expenses for enlargement. These are 'ring-fenced', which means that 'amounts available for accession can only be used in order to cover expenditure arising as a direct consequence of enlargement' (Presidency Conclusions, No. 11). The European Council considered this to be 'the total cost of enlargement for each of the years 2002-2006' (Presidency Conclusions, No. 11).

Table 3.1   Expenditure for Enlargement 2002-2006

Million €

Year	2002	2003	2004	2005	2006
Total	6.450	9.030	11.610	14.200	16.780
Agriculture	1.600	2.030	2.450	2.930	3.400
Struct. operations	3.750	5.830	7.920	10.000	12.080
Int. policies	730	760	790	820	850
Administration	70	410	450	450	450

At this point it is safe to assume that there will be no country formally acceding in 2002. Amounts not used in the year(s) before the enlargement takes place will not be available for enlargement in later years, nor for any other purpose. It is a matter of speculation whether this can and will be adjusted in case of need.

A comparison with expenditures for the EU-15 helps to assess the figures of the financial framework. There is a very pronounced discrepancy between the appropriations for *agriculture* in the EU-15 and the appropriations for new Member States. The framework is based on the assumption that the number of new Member States at that time will be six, namely Cyprus, Czech Republic, Estonia, Hungary, Slovenia and Poland.

Table 3.2   Appropriations for Agriculture in EU-15 and Accession Countries

Million €

Year	2002	2003	2004	2005	2006
EU-15	43.900	43.770	42.760	41.930	41.660
6 acc states	1.600	2.030	2.450	2930	3.400

The expenditures in agriculture for new Member States will rise from 2002 to 2006, but even for the last year of the financial perspective the appropriations for the six new Member States are just over 8 per cent of the foreseen appropriation for the EU-15. This was based on the assumption that farmers in the new Member States would not receive compensatory payments.

In view of the ongoing CAP reform and in view of the uncertainty of the rules applicable in the new Member States, it is very difficult to assess the real cost of enlargement in the area of agricultural policy. Conservative forecasts assume a minimum expenditure in the range of at least 20 per cent of expenditure for the EU-15; this includes compensatory payments.[5]

The discrepancy between expenditures in old and new Member States is not quite as pronounced in the area of *structural operations*, at least not for the later years. In looking at these figures, it is necessary to keep in mind that they are based on the rules governing these structural expenditures for the EU-15 through the year 2006. Changes which might be decided on for future periods after accession are a different matter; they will be discussed below.

Table 3.3   Appropriations for Structural Operations in EU-15 and Accession Countries

Million €

Year	2002	2003	2004	2005	2006
EU-15	30.865	30.285	29.595	29.595	29.570
6 acc states	3.750	5.830	7.920	10.000	12.080

The expenditures for new Member States amount to over 33 per cent of the expenditures for the EU-15 in 2005 and slightly over 41 per cent in

2006 (not taking into account changes which might be the consequence of the mid-term review of the Cohesion Fund in 2003).

The figures for the EU-15 also include expenditures for Objective 2 and Objective 3 and Community Initiatives. A comparison between expenditures for the EU-15 and the accession states should concentrate on the resources allocated for *Objective 1* and the *Cohesion Fund.*

Table 3.4   Resources Allocated to Objective 1 and Cohesion Fund

Million €

Year	2002	2003	2004	2005	2006
EU-15: Obj1, CF	22.305	21.900	21.390	21.390	21.092
6 acc states	3.750	5.830	7.920	10.000	12.080

The Objective-1-regions of the EU-15 for the years 2000-2006 comprise a population of 83 million. Objective-1-regions of the six applicant states of the 'first group' would have a population of about 60 million (total territory of the six states minus Prague and minus Cyprus, unless separate regions were formed in Cyprus, some of which might come under the 75 per cent ceiling).

In assessing these figures we should keep in mind that payments from the structural operations are limited to four per cent of national GDP, as mentioned above. The national GDP of the five CEE countries of this group amounted to a total of 211,011 million ECU (four per cent equals 8.440 million) in the period 1995-1997 (average). For each of the five countries the four per cent ceiling would have been as follows:

- Czech Republic: 1,758 million;
- Estonia: 137 million;
- Hungary: 1,467 million;
- Poland: 4,474 million;
- Slovenia: 604 million.

The GDP in these countries is likely to be higher in the reference year for expenditures in 2006. Cyprus must be included, since this country would get transfers from the structural operations even though it does not qualify, or does not qualify totally as an Objective-1-region.

It is hard to predict the GDP of these countries in the future reference year for expenditures in 2006.[6] The amount provided for in the financial framework (€ 12.080 million) might just about correspond to the four per cent ceiling, if all of these countries are taken together, i.e. more could not be allocated under the existing rules. The amount provided would very likely not reach this ceiling if a total of ten countries joined before 2006, i.e. the four per cent rule would allow a higher expenditure.

The framework for the present period fixes a clear overall ceiling on the budgetary resources available and on the appropriations for the different objectives. It can be assumed that the EU will not raise the overall ceiling laid down for each year. Changes between the different headings might be another matter. Still, even such changes will not be easy: For one, there is the distribution of funds between the EU-15-Member States and new Member States. Then there is the agricultural guideline which according to the Berlin European Council will remain unchanged. However, it is to be reviewed before the first enlargement (Presidency Conclusions, No. 19).

As mentioned before, the EU might have to mobilize some of the resources allotted for the years beginning at 2002 and use them for the new Member States once accession has become effective at some later time before 2006.

Future Financial Frameworks

The impact of enlargement will be felt fully in the next financial period beginning in 2007. Thus it is of importance whether enlargement at least by a first group of countries happens before that date – and in time for the negotiation of this new framework.

It is still too early to assess the volume of the new financial framework. Much depends on the development of the GNP in the Member States because the overall own resources ceiling is expressed as a percentage of the GNP.[7]

The new financial framework cannot differentiate any more between old and new Member States. All States which are EU Members as of 1st January 2007 have to be treated simply as Member States and without distinction as to the date on which they joined the Union.

The new financial framework might also have a separate chapter for enlargement. This can only concern those applicant states which will not have acceded before 2007 and which might become Member States at a later date during that period.

As past experience shows, fixing the financial framework is a political decision in its own right, i.e. the budgetary allocations are not simply a consequence of separate decisions in the different areas of EU policy. Those decisions, in their turn, have to respect the financial framework once this has been agreed upon. This is especially pronounced at times when Member States themselves have to make great efforts to cut public expenses in order to consolidate their budgets.

In setting the framework, the European Council takes into account what kind of policies should be pursued during the next period. The overriding purpose of the framework is, however, to provide a secure base for planning budgetary expenditure on the EU level and for the national contributions. Allocations for the different areas of EU policy are, with few exceptions, set without determining in advance every detail of what the policies will look like.[8]

This is illustrated by the treatment of the expenses under Heading 1 (Agriculture) by the Berlin European Council. The Council fixed ceilings on agricultural expenses for each year. In addition it provided that an 'average level of € 40.5 billion plus € 14 billion over the period for rural development as well as veterinary and plant health measures' (Presidency Conclusions, No. 21) should not be exceeded. It requested the Council and the Commission to pursue additional savings to ensure that expenditures will not overshoot this average annual expenditure.

It becomes quite clear from this that in the view of the European Council the agricultural policy and its further development during this period will have to respect the budgetary allocations under this financial framework – and that the policies (and not the allocations!) have to be adjusted if the expenses overshoot the allocations.

Cohesion Policy

Cohesion policy is, next to agriculture, the area in which the EU will be presented with the greatest challenges in applying its policies to the new Member States.

New Members are likely to be admitted to the Union before their economic development has caught up with the EU's average or even with the average of the so-called 'Cohesion countries' of the present EU. The Copenhagen criteria are not interpreted in a way as to require a certain GDP/capita in the countries which join.

According to the latest publication by Eurostat (News Release 48, 2000),[9] the GDP/capita in the Czech Republic and Slovenia – the two most developed of the CEE countries – reached 64 and 66 per cent of the EU-15 average respectively (average for the years 1995-1997 in PPS). The average GDP/capita in the other accession states from Central and Eastern Europe was still markedly lower. In per cent of EU-15 average:

- Hungary: 47 per cent;
- Slovakia: 44 per cent;
- Estonia: 34 per cent;
- Poland: 34 per cent;
- Romania: 32 per cent;
- Lithuania: 29 per cent;
- Latvia: 25 per cent;
- Bulgaria: 25 per cent.

In addition, we must keep in mind that there are important regional differences within the accession countries – differences which might even widen in the process of further economic development of these countries. According to Eurostat, in the Czech Republic the spread is between 49 and 119 per cent of average EU-15 GDP/capita, and in Poland between 24 and 49 per cent. There are important regional differences also in the present Member States.

Even taking into account that the gap between the accession countries and the Member States will narrow before accession, one may assume that the accession countries will still be at or around the level of the least developed of the present Member States, i.e. Greece, Portugal, and certain regions in Spain, when they join the EU (Gutachten of the Friedrich-Ebert Stiftung, 2000). There can be no doubt, therefore, that the inclusion of these countries will have a major impact on the EU cohesion policy.

Objectives of Cohesion Policy

According to Art. 2 of the Treaty on European Union it is one of the objectives of the Union: 'to promote economic and social progress and a high level of employment and to achieve balanced and sustainable development, in particular through (...) the strengthening of economic and social cohesion'. Art. 2 of the Treaty establishing the European Community notes that 'economic and social cohesion and solidarity among Member

States' is one of the tasks of the Community.[10] For both Treaties, the principle of subsidiarity applies also to economic and social cohesion policies.

Art. 158 of the EC-Treaty spells out the objective of economic and social cohesion in a somewhat different wording: 'In order to promote its overall harmonious development, the Community shall develop and pursue its actions leading to the strengthening of its economic and social cohesion. The Community should aim at reducing disparities between the levels of development of the various regions and the backwardness of the least favoured regions or islands, including rural areas.'

The wording of Art. 158 makes it quite clear that disparities should be 'reduced', but not necessarily altogether compensated. That is a realistic approach. Disparities between regions will remain and cannot be levelled entirely. One also has to keep in mind that 'cohesion' is not limited to the regional aspect, as is demonstrated by the horizontal measures of the social fund, for instance.

Art. 159 EC-Treaty adds two important elements: First, it makes clear that cohesion is foremost a responsibility of Member States: all the contributions by the EC will not have the desired effect if Member States do not conduct their policies in a way that is conducive to cohesion. Secondly, the provision establishes the obligation for the EC to take into account the objectives of Art. 158, i.e. of economic and social cohesion, in the formulation and implementation of all of the Community's policies and actions and the implementation of the internal market.

As a general principle the aim of economic and social cohesion permeates all of the EU policies, to the extent that they lend themselves to taking cohesion into account. The aspect of cohesion is of particular relevance to policies involving the transfer of money, which is not surprising. These are primarily the structural measures and the Common Agricultural Policy, but the net contributions/receipts of Member States also merit consideration in this context.

Structural Measures

Accession will have a profound effect on the present cohesion policies of the EU. Apart from the financial transfers and their re-orientation that the accession may require, there is also a more basic problem which the EU will have to deal with, namely, the question of determination of eligibility. Presently a relative scale based on GDP/capita in relation to the EU average is used. Eligible for Cohesion Fund transfers are Member States with an average GDP/capita that is less than 90 per cent of the EU average. The threshold for Objective-1-regions under the Structural Funds is 75 per cent of the EU average.[11]

Apart from the Cohesion Fund the following discussion can be limited to the Ojective-1-regions because 69.7 per cent of the resources available for Structural Funds are distributed in these regions (cf. Decision of the Berlin European Council for the period 2000-2006), and the territory of the new Member States will be mostly classified as Objective-1-regions.

Including the new Member States into such a scale would move the average GDP/capita in the enlarged EU downward – and with it the 90 and 75 per cent thresholds, as an obvious result (Friedrich-Ebert Stiftung, 2000). The GDP/capita in Slovenia, which amounts to 66 per cent of EU-15 average, would rise to 76 per cent of EU average if the 10 CEE countries were included. For other countries the figures as reported by Eurostat (News Release 48/2000) are as follows:

Table 3.5   Shifts in GDP/Capita in Relation to the EU-Average

BUL:	from 25 per cent to 29 per cent;
CZ:	from 64 per cent to 74 per cent;
EST:	from 34 per cent to 39 per cent;
HUNG:	from 47 per cent to 54 per cent;
LITH:	from 29 per cent to 33 per cent;
LAT:	from 29 per cent to 33 per cent;
PL:	from 34 per cent to 39 per cent;
ROM:	from 32 per cent to 37 per cent;
SLOVAKIA:	from 44 per cent to 50 per cent.

Such a realignment of the 'cohesion-scale' would seem inevitable. It would exclude some of the present Objective-1-regions from transfers under this heading, but most likely not from transfers under the Cohesion Fund, with the exception of Ireland. This realignment is likely to present the EU with political problems.

Beyond the realignment, there is the more fundamental question, however, of what 'cohesion' means and how to 'strengthen' it: countries or regions which are already relatively developed are likely to close the gap faster than those at the bottom-end of the scale. This will increase the gap between the relatively more developed and the poorest regions. The new Member States are most likely to find themselves among the latter.

This effect will make it necessary, in any case, to apply measures of cohesion in a way that takes the diversity of the less developed regions fully into account. For this, one may refer to the conclusion of the Berlin European Council which includes the following criteria for the allocation of resources: 'Eligible population, regional prosperity, national prosperity and the severity of structural problems, especially the level of unemployment' (Presidency Conclusions, No. 45). In addition, an 'appropriate balance will be struck between regional and national prosperity' (Presidency Conclusions, No. 45).

Present Situation In view of the structural policies of the EU in the past, the following points can be made:

- The structural policies are connected with a considerable transfer of resources.
- The structural policies assisted the less developed Member States to a considerable extent in their economic and social development.
- Not all of the Member States receiving net transfers made use of these funds to the same extent, i.e. the positive effect in the Member States and regions receiving these funds differed considerably from Member State to Member State, as well as from region to region.
- The net transfer did not follow strictly the criteria set by the EU itself but it was also determined by political expediencies, interfering with the overall aim of cohesion.

The conclusions of the Berlin European Council on the Agenda 2000 are a good example to demonstrate the last point. Rather than taking fully into account the changed situation of Member States and regions, the European Council provided considerable funds for regions which no longer meet the criteria for support under the different objectives (valid through 2005 and 2006, respectively).[12]

The European Council provided the following reasons for such transitional payments: 'Adequate transitional support (...) are (sic) an essential counterpart to greater concentration of structural funds, so as to underpin the results secured by structural assistance in ex-Objective-1-regions' (Presidency Conclusions, No. 42).

This is a misleading argument. In fact, the opposite is true. Continuing such a considerable support for Member States and regions which are no longer eligible under the standard rules established for structural assistance runs counter to the EU objective of strengthening cohesion. I am not sure whether the European Council was aware of this fact.

It is also questionable whether the reference period chosen for determining the eligibility as Objective-1-regions – the average of the last three years for which statistics were available in March 1999 – is really justified. After all, this determines the assistance for the seven full years to come, namely from 2000 to 2006. One may assume that a number of these regions will – as in the past period – exceed the 75 per cent criterion during this period, and they will continue to receive these transfers. One can only hope that it will not be thought necessary at the end of this period to repeat similar 'transitional' measures for these regions.

After Enlargement For the structural policies after accession – I refer here to the period after the present financial period, i.e. to the period beginning in 2007, and assume that accession by at least some of the CEE states will occur beforehand – several alternatives present themselves.

In detail, the solutions and their consequences will depend on how many and which countries will become new Member States first, and on when the rest will follow. The budgetary framework will also be a determining factor. It is unlikely that the present ceiling for structural measures of 0.46 per cent of the GNP of the EU (including the new Member States after accession) will be raised (Communication of European Commission of 16 July 1997 on Agenda 2000). The actual amounts available within this framework depend on the development of the GNP in the old and new Member States.

These are some possible solutions:

- Leaving the acquis unchanged, i.e. applying the thresholds of 90 and 75 per cent for eligibility for the Cohesion Fund and Objective 1 regions respectively, calculated on the basis of the present 15 and the new Member States;
- Leaving the acquis unchanged, but retaining the EU 15 as a basis for calculating the thresholds for all Member States;
- Maintaining the acquis for the present Member States (as was done for the financial period until 2006) and establishing special rules for the new Member States;

- Changing the acquis by adapting the thresholds or introducing other measures.

The first alternative would be the most obvious solution. However, I fear that the EU will want to change the acquis at least for the period of the next financial framework. In the longer run one will have to return to this formula, however.

The second option – using the average GDP/capita of the 'old' 15 Member States as a basis for calculating the thresholds – would be difficult to justify. This was in part attempted following the German unification, based on the argument that there were no reliable statistics covering the reference period for the new *Laender*. However true this might have been, it was obvious to everybody concerned that the GDP/capita of the new *Laender* was below the 75 per cent threshold of EU average (in fact, their regions were included as Objective-1-regions on that basis alone!). The motives of the argument had of course nothing to do with statistics but was politically motivated in order to secure resource transfer to regions which otherwise would be above the thresholds.

Maintaining the acquis for the actual Member States and establishing new rules for the new Member States – the third option above – would not be consistent with the objective of economic and social cohesion. This would make it even more difficult for the new Member States to 'catch up' and would thus further question the effectiveness of the cohesion policies. More importantly, such a policy would not be compatible with the rule of equal treatment.

The EU maintains vis-à-vis the applicant countries that they have to accept the acquis of the EU unchanged. One would think, therefore, that the EU itself would want to maintain its acquis also for the structural measures, including the present thresholds calculated on the basis of the GDP/head of all the States which are Members at any given time. It is important to recall in this connection that the *rules* governing the expenditure under the Structural Funds and the Cohesion Fund constitute the acquis and *not* the transfers as such. This means that in the application of the acquis the list of regions falling under Objective 1 must be adjusted accordingly.

With regard to the fourth proposal, especially two intermediate solutions seem possible:

- New transitional measures for Member States/regions which would find themselves above the thresholds if the acquis were applied unchanged (i.e. by including the new Member States and their regions into the calculation).

• The creation of new (sub-)categories of Member States and/or regions (while including the new Member States into the relative scale of GDP/capita). This could be a category of Objective-1a-regions somewhere above 75 per cent of EU average, with reduced support.

Whatever solution is finally chosen: the relative regional and national prosperity will have to be taken into account in such a way as to favour sufficiently the Member States and regions at the lower end of the prosperity scale. Otherwise, the EU measures for cohesion would support the relatively more prosperous States and regions to the detriment of the less prosperous ones. There is no way around this: if the EU wants to conduct an all-inclusive cohesion policy, some of the States or regions that are favoured under the present regime will have to forego resource transfers in the future. This is also the view of the Commission as expressed in its Communication of 16 July 1997 on Agenda 2000.

The application of the ceiling of 4 per cent of national GDP for transfers from structural operations (see above) might lead to problems in this context and might require the application of compensatory measures in order to maintain the relative scale of support for Member States and/or regions.

Net Receipts/Contributions

One may – and I believe one should – call attention to coherence also when the net contribution or receipt of Member States is being assessed. Art. 158 refers to the 'overall harmonious development' in the EC. Net contributions/receipts of certain Member States might run counter to this aim.

Nobody would dispute that Member States with a GDP/capita below 75 per cent of EU average should not be net contributors. Such a result would not be compatible with the aim of economic and social cohesion – whatever the contributions from the structural funds may be in that Member State. The principle of cohesion can, of course, not provide exact criteria for the amount of net contribution or net receipt by or to any Member State. But, if the scale of net contributions per head of population deviates widely from the position of the individual Member State on the scale of GDP/capita in the EU, the principle of economic and social cohesion will require adjustments.

I do not believe that the scale of net contributions/receipts which is the effective result of the Berlin European Council reflects this consideration sufficiently. It is beyond the scope of this chapter to elaborate this aspect in detail. I would just like to point out a few figures from past years in order to exemplify the problem. These are taken from the Court of Auditors report of 1998 and the Commission Communication on the execution of the budget for 1999 (SEC, 2000, 340).

* France and the Netherlands had about the same GDP/capita in 1998. The net contribution amounted to 29,4 ECU/capita in the case of France, yet to 195,3 ECU/capita in the case of the Netherlands (1999 figures)!
* On the receiver side, there is a discrepancy, for instance, between Ireland, on the one hand, and Greece, on the other: The GDP/capita in Ireland in 1998 was 80 per cent (in ECU) or 50 per cent (in PPS) higher than in Greece. As a net receiver, Ireland received 577 ECU/capita whereas Greece received only 435 ECU/capita in 1999!

Such discrepancies (positive or negative) in net transfers from the EC budget seem to run counter to cohesion. Such problems might also arise after the accession of new Member States.

There seems to be a general agreement that it would be unacceptable for any of the new Member States to become a net contributor. This does not answer all the questions, however. Since the GDP/capita in all of these states is and might remain lower till than the GDP/head in all of the present Member States for some time after accession, the principle of economic and social cohesion requires that their net receipts per head are in relative terms comparable to those of the present Member States with the lowest GDP/capital. Within the EU financial framework, this will require a mutual adjustment of resource allocation.

Agricultural Policy

There is no other area of EU policy which will present the accession negotiations and the future policy making with as many and as difficult problems as the Common Agricultural Policy (CAP).

There are many reasons for this:

- The structure of agriculture within the present EU and in most of the applicant states;
- The often fundamentally opposed interests of the EU Member States, and also, one may assume, of the applicant states;
- The present CAP, which in itself makes it difficult to admit new Member States with a potentially high agricultural production;
- The rules of the CAP and related issues, which are often extremely complicated and difficult to apply, especially by any 'newcomer';
- The transfer of considerable resources from the EC budget to the different Member States, a consequence of the CAP.

One of the major problems of the highly developed and sophisticated CAP is that farmers do not make decisions as dictated by the market but as dictated by the rules of the CAP. This also has the effect that the future development of agriculture, including its structure, is determined largely by the decisions on the CAP and not by market developments. Often enough, these decisions form a patchwork based on the different interests of Member States (or: their assumed interests). There is no general concept of an EU agriculture on which the Member States have agreed on, nor on how it should develop. Such a general concept might not even be possible in view of the diversity of conditions and interests within the EU.

The principles laid down by the Berlin European Council do not provide much of an indication of what the CAP will look like in the future:

> that agriculture is multifunctional, sustainable, competitive, and spread throughout Europe, including regions with specific problems, that it is capable of maintaining the countryside, conserving nature and making a key contribution to the vitality of rural life, and that it responds to consumer concerns and demands as regards food quality and safety, environmental protection and the safeguarding of animal welfare (Presidency Conclusions, No. 20).

Apart from the reference to 'competitiveness', there is no mention of the 'market', let alone a reference to Art 4 EC-Treaty on economic policy (of which agricultural policy, after all, is a part) and its guiding principles.

Reform of the CAP

There is general agreement that the CAP needs further reform. In preparation for the Berlin European Council, reform measures were discussed extensively, and the Agricultural Council reached an agreement on a number of measures in March, 1999.

The European Council however did not push the reform as far as it could or had been hoped: the intervention price reductions for cereals were less than had been proposed; the dairy reform will enter into force only as from the 2005/2006 marketing year; and compensatory payments were not made degressive – the financing of such payments in part through national budgets was not agreed on. The European Council was nevertheless successful in obtaining agreement for a reduction of the total expenditure for agriculture, excluding rural development and veterinary matters, to an average of € 40.5 billion annually in the 2000-2006 period. It expressed the opinion that the reform of the CAP could be implemented within this framework, and it invited the Commission to submit a report in 2002 accompanied, if necessary, by 'appropriate proposals' (Presidency Conclusions, No. 22).

As pointed out above, the allocations for agriculture under the heading 'Enlargement' in the financial framework 2000-2006 will not be sufficient to meet the requirements from extending the CAP – whatever the rules might be in detail – into six or even ten to twelve new Member States. The agricultural guideline calculated on the basis of 21 or more Member States will not be adequate to accommodate these additional expenses on the basis of the present CAP because the GDP of the acceding states is relatively low. The EU will thus be faced with the choice between reforming the CAP drastically or raising the agricultural guideline.

A drastic reform would constitute the optimal solution under economic and financial aspects. One can only hope that the EU will make use of the opportunity that enlargement presents to develop a more sensible agricultural policy. The results of the Berlin European Council are not too encouraging in this respect, however.

The alternative would be to raise the agricultural guideline. This guideline proved to be an effective tool for curtailing agricultural expenses and their further growth. If this guideline were changed, there would be the real danger that it would lose its 'braking' function. The pressure on reform would diminish, and interested Member States would use this adjustment to extend the guideline even further.

There are some principles which guide the reform, and which will remain important for its future development. Most likely, the tendency away from price fixing and intervention and/or quota systems and away from the corresponding protection against imports will continue. Up till now, measures and systems to guarantee prices were replaced by compensatory or other direct payments. It is doubtful to what extent such payments in their present form can be maintained under the WTO rules in the future. More generally, developments in the WTO will have a major – and generally salutary – influence on the reform.

Further reform is of great relevance for accession. The prospect of accession and its consequences for the CAP should be one of the guiding factors for the measures to be taken. If the reform is postponed till after accession, or till after the Accession Treaties are signed, any reform is likely to become much more difficult. In addition, it would be almost impossible to close the negotiating chapter on agriculture as long as it is unclear what the reform will look like in detail.

It has to be kept in mind in this context that for a number of countries the next big factor of relevance for cohesion is the Common Agricultural Policy. For instance, in Ireland the CAP transfers exceed the transfers from the structural funds, whereas in Portugal they reach only 20 per cent of structural funds contributions. This contributes to Portugal's weak position in the net transfers per head of population. It is of interest in this context that the Berlin European Council decided to double the existing MGA for durum wheat in Portugal from 59,000 to 118,000 ha (Presidency Conclusions, No. 22).

Accession Negotiations

As in other sections, I do not propose to discuss the particular problems of the accession negotiations. In agriculture, however, a number of these questions will also be important for the formulation of EU agricultural policy in the future.

I propose to leave aside minor questions even though they might also throw a light on the problems that the CAP could face in the future: among these I count the request of Slovenia that an exception be granted in favour of a Slovenian wine specialty that is a blend of white and red wines, or the request of Cyprus concerning the minimum length of bananas. Even these minor issues indicate what kind of questions might arise concerning the future CAP and its numerous and often very detailed rules.

Of greater and more direct importance are questions like ceilings for certain kinds of cattle, for cereals and oils seeds or the milk and sugar quotas and their allotment between the (old and new) Member States.

In reading the Position Papers of the applicant countries, on the one hand, and the Draft Common positions of the EU, on the other hand, it becomes clear – not surprisingly – that the positions on these questions are far from being identical: the applicant countries seek quotas or ceilings as high as possible, whereas the EU wants to keep the quotas and ceilings for the new Members down. The EU refers to the actual production in the applicant countries,[13] which is – conveniently enough – lower than the production at other times during the past and much lower than the production *potential* in these countries.

An example of the 'games' which are played in this context is the practice in an applicant state – or, so the EU Commission assumes – of 'artificially' raising production in a quota-governed product in order to provide arguments for the allotment of higher quotas during the accession negotiations. In my opinion, this proves that this country has already understood the mechanisms and 'opportunities' of the CAP.

Future Policy

I would like to examine the influence of accession on the CAP by selecting a few sectors and problems areas of that policy. There is no area of EU policy which has been examined and discussed so heavily as the CAP. For that reason, let me just contribute a few considerations that will complete the general picture of the impact of accession on EU policies.

I will not include a discussion on the market regulations for milk and dairy products and for sugar. It is likely (and to be hoped!) that these regulations will be changed substantially during the CAP reform. According to the Berlin European Council, the dairy reform is to become effective as from the 2005/2006 marketing year. There is a general presumption that the quota system, which has been criticised heavily since its beginning, will have to end at some time.

Direct Payments The EU agricultural policy relies less and less on fixing prices but has replaced price support more and more by direct payments to the farmers. These payments are product-related. The rules for such compensatory payments vary from product to product.

This tendency of the reform is likely to continue, also under the influence of the WTO rules. As mentioned above, it is disputable whether these payments in their present form can be maintained under WTO rules in the future.

The position of the EU, as reflected by the Berlin European Council, seemed to be originally that direct payments were not necessary in the new Member States, as these payments were meant to offset the reductions of guaranteed prices. Since prices were lower for these products in the applicant states it was argued that it should not be necessary to offer such compensatory payments to farmers in these countries.

This position was, at least to a certain extent, defensible as long as such direct payments were to decrease over time and to end at a certain date. If that had been decided, it would also have facilitated the negotiations, because such payments would have reached a low level at the time of accession already and the remaining time before their total abolition would have been relatively short.

The Berlin European Council, however, did not adopt the concept of decreasing these payments and limiting them in time. It is difficult to predict whether this rule might be introduced into the CAP at a later time.

In the meantime, there is a growing awareness in the EU that its original position (no direct payments) cannot be maintained.[14] A question might remain, however, as to the extent of these payments in the new Member States. It has been suggested that such payments should be replaced in the new Member States, at least partially, by assistance for the restructuring of agriculture. There is much to be said for such an approach. This also holds true for the present Member States.

When assessing the extension of such payments into new Member States we must look at different market regulations. In assessing them – and their extension into new Member States – we should keep in mind that such payments are in a number of areas dependent on supplementary actions by the recipients, such as the set-aside.

Beef and Veal This market regulation and the payments provided therein are quite varied. The payments serve different purposes, and they have to be assessed individually. Without going into all the details of this regulation, the following payments might be mentioned: special premium for male bovine animals, within certain limits and within a regional ceiling; additional premium (deseasonalisation premium) for steers slaughtered between the 15th and the 23rd week; premium for maintaining suckler cows, within individual ceilings which have to respect a national ceiling; slaughter premium for (all) bovine animals within a national ceiling; extensification payment under certain conditions concerning the stocking density; headage payments and area payments. In addition, a 'basic price' is set for carcasses of male bovine animals. Private storage aid may be granted if the market price remains less than 103 per cent of the basic price. Beginning in July 2002 public intervention shall be opened if the market price in a Member State or in a region falls short of a certain amount.

In assessing this regulation and the impact of enlargement on it, I should like to present the following considerations.[15]

As long as the price-related measures are applicable, it appears impossible either to vary the prices (basic price, intervention price) between old and new Member States or not to apply the storage aid or public intervention which are connected with these prices. Within the internal market, including the free movement of meat, it is unthinkable not to apply price-supporting measures in some Member States/regions. This would have an immediate effect on the prices in the other regions.

It is likely, though, that these prices and the related measures will lose some of their importance in the future, since the objectives of the reform of CAP are to move away from price support and determine payments to the producers independent of market prices.

As far as the payments are concerned, the extensification payment and the area payments must also be provided in the new Member States. These payments serve a purpose beyond the mere meat production: they are meant to support (essentially for environmental reasons) extensive agriculture and to maintain permanent pasture.

The same holds true, albeit for other reasons, for the deseasonalisation premium. This is meant to take pressure off the market. This rationale requires that it be applied in the whole Union. Possibly differentiations could be introduced according to the situation on different segments of the market. This would not justify a general differentiation between old and new Member States, however.

It is only the more general premiums that lend themselves to a discussion about whether and to what extent they should be applied in new Member States.

As in other areas of the CAP, the question of fair competition between producers from different areas of the EU will have to be examined: in all of the EU there will be a more or less homogeneous market with comparable prices. All producers will be able to sell their beef and veal for these prices wherever their production might be located.

Under the present market regulation the producers in the old Member States receive price support in the form of these premiums. There is no differentiation as to the level of prosperity in the different Member States or regions. This puts those farmers at a competitive advantage over producers in new Member States. One cannot assume generally that producers in the new Member States will produce the meat at a lower cost after accession than all of their competitors from the old Member States.

Sheep Meat and Goat Meat The regulation provides for the fixing of a basic price for fresh and chilled sheep carcasses. An average weighted price is fixed weekly for each Member State. In order to offset losses a premium is paid to the producer if the average market price is lower than the basic price. For each producer an individual limit is introduced.

This regulation states expressly that the 'expenditure (...) shall be deemed to form part of intervention for the purpose of stabilizing agricultural markets' (Council Regulation (EC) No. 2467/98 1998). It seems inconceivable that new Member States would not be included in this systems of premiums. The fixed basic price cannot be lower in certain Member States than it is in others. This would not be compatible with the rules of the internal market.

Let us assume that farmers in the old Member States produce a high number of sheep carcasses which are sold EU-wide. This might bring the EU market prices down if the production exceeds the consumption in those countries. The farmers in the 'old' Member States would receive a premium based on these lower market prices. Their excess production might be sold in the new Member States, and the market price would go down there, too. If premium payments were differentiated under such conditions, the EU regulation could not achieve its purpose to 'stabilise the markets'.

The regulation, as it stands today, refers in several of its provisions to the 'Community markets', 'throughout the Community', or the 'Community as a whole'. This could not be maintained if the whole system of paying premiums were limited to the old Member States. If one did not include the new Member States into the system of this regulation, one would create, in effect, two different markets within the Union.

As far as goat meat is concerned, the premium is limited to certain areas, i.e. it is not paid to goat meat producers in all of the Community. This differentiation has something to do with selecting the traditional areas of goat meat production. It cannot be regarded as a precedence for limiting sheep meat premiums to the old Member States.

The only question which might remain is whether the market regulation can be changed in such a way as to vary the amount of the premium paid. This will be very difficult as long as the premium within the EU is the same regardless of where the meat is produced.

There is also the danger that such a differentiation puts the producers in the new Member States at a competitive disadvantages because they would have to compete with meat, the production of which was subsidised by the full premiums.

Cereals and Oilseed The Berlin European Council decided that the area payment for cereals shall apply also to oilseeds from 2002/2003 onward. The reference for these payments is 'aid per ton multiplied by historical reference yields for cereals' (Presidency Conclusions, No. 22). The European Council also decided that the 'intervention price for cereals shall be reduced by 15 per cent' and the area price shall be increased 'to 63 €/t (multiplied by the historical regional reference yield for cereals)' (Presidency Conclusions, No. 22).

The payments are linked to the so-called set-aside which is meant to reduce over-all production. As in other cases, these area payments lend themselves to a discussion if and to what extent they should be applied in new Member States. The question of fair competition will also have to be addressed in this context.

Enlargement and Further CAP Reform

The CAP reform will continue. There can be no doubt about its general direction: lowering or abolishing price guarantees and quota systems and providing direct payment to farmers instead.

This tendency of the reform process needs to be strongly supported. Some major problems still need to be resolved, however:

- Such payments should not be permanent under the heading of agricultural policy. Attempts at adopting such a principle failed at the Berlin European Council. The idea was right. It should remain part of the reform package.
- The payments need to be dissociated from production. This will (most likely) be necessary under WTO rules.
- Permanent support should be extended only to certain farmers under social aspects. This would fall into the responsibility of Member States (social policy, not agricultural policy).
- None of the measures under the CAP should be so designed as to hinder, delay, or even counteract the still necessary substantial restructuring of agriculture (within the EU-15 and the new Member States).

In further reforming the CAP, the EU must take into account the consequences of enlargement, including the budgetary consequences. The reform should be concluded before the accession of the first new Member States. This is no argument for delaying accession, but for speeding up the reform of the CAP. There are several reasons for this sequence (apart from inner-EU concerns and WTO-related considerations):

- Accession and accession negotiations might be very much more complicated if the reform were not completed until some time after accession. A framework for the inclusion of the accession countries into the present CAP would have to be negotiated and they would have to be included into the CAP, even though some basic data of that policy would be changed again in the very near future after accession.
- The reform has to come in any case in time for the negotiations on the next financial framework.
- Any reform will be much more difficult even after the first round of accessions. Since the new Member States have (with significant differences between them) important agricultural sectors which in some countries still need a major restructuring, it is likely that the new Member States (as most old ones always did) will look at any reform under the aspects of what own farmers get and what the financial return for the country will be.

The prospect of accession should strongly support the general tendency of the reform and some of its more concrete objectives. The EU should not live under the illusion that it can exclude new Member States permanently from the CAP's rich benefits for farmers (and the benefits that follow from it for some national budgets).

If accession is not taken into account sufficiently, the budgetary burdens might become unsupportable. This should also be of concern to present Member States which are net recipients in terms of agricultural expenses. In view of the size of the agricultural sector in several of the new Member States, the positive net balance of these present Member States could worsen drastically. Finally, the (very necessary) restructuring of agriculture in the new Member States would be slowed down. As long as farmers in the accession states can hope for sizable transfers under the CAP after accession they might be reluctant to join restructuring measures.

Competition Policy

Fair competition is of prime importance for the EU's internal market. The chapter on competition in the EC-Treaty testifies to that. Accession by new Member States cannot compromise this principle of fair competition. Still, in its application it will be necessary to take particularities in the new Member States into account.

There are also a number of provisions in the secondary EU law which are motivated exclusively or in part by considerations of ensuring fair competition within the internal market. Fair competition might become an issue in connection with this legislation and its application in new Member States, for instance in the area of environmental law or social policy.

The question of fair competition will be of importance for the accession negotiations where it has been raised already. It will remain an issue also for the development of the EU's policy after accession. This will especially hold true if the EU policies should take diversity into account to a greater extent than has been the case till now.

In assessing 'fair competition' and the impact of the different rules on the competitive situation in the different Member States (old and new) and different regions, one has to take into account that situations or conditions for business might also differ. It might very well be that uniform rules which are meant to ensure fair competition contribute to 'unfairness' if they do not treat different situations differently. The *Standortfaktoren* (local conditions) are different in the individual Member States and in different regions of Member States.

To the extent that Member States themselves can influence these conditions (for instance, taxation) it can be left to their responsibility to adjust these conditions so as not to harm the competitiveness of their industry. Infrastructure is also largely the Member States' responsibility but its improvement is costly and may take a long time.

The competitive position of industry is also influenced by agreements on wages and working conditions. There are natural conditions (such as the weather) which cannot be influenced by the Member State concerned or by anybody else. And there are social conditions (in the widest sense) which can be influenced by Member States, social partners, or others but which do not lend themselves to rapid change, such as the level of education.

It seems quite obvious that an EU-wide minimum wage at a level that would have any meaning in Germany would be insupportable for industry in some of the present Member States and very likely also in the new ones. This is to show that homogeneous rules do not always correspond to the principle of 'fair competition'.

It is difficult to strike the right balance between taking into account diversity and laying down common rules with a view to ensuring fair competition. It is very likely that Member States, regions, and business will have different opinions about it in every case: Some will argue in favor of homogeneous rules in order to ensure fair competition, possibly also having in mind that competitive pressure would thus be reduced. Others will argue for taking diversity into account in order to give their industry a competitive chance on the market.

This conflict of interests is clearly apparent in the area of state aid which I discuss below. But it is also of great importance when establishing social standards or standards for the protection of the environment; I will turn to these questions under the sections on social policy and the environment.

Control of State Aid

National aid to industry which might distort competition in the internal market is, in principle, prohibited. It might, however, be authorised by the EU Commission under certain conditions. The European Court of Justice recognises that the EU Commission has a broad discretionary power in approving national aid.

In spite of the general prohibition of state aid, such aid plays an important role in Member States' policies, as well as in the policy of the EU itself, especially if one includes agriculture.

It certainly can be debated if and to what extent such aid is good economic policy, and if it produces the effects intended. These questions do not have to be discussed in this context. Such doubts exist first and foremost concerning aid in the more developed Member States and regions. One may refer to such cases as the coal production in the Ruhr valley, or the restructuring of Air France.

Beyond that, we can base our discussion on the accepted EU policy, which includes, among others, regional aid, aid to small and medium sized enterprises, and aid for restructuring enterprises in difficulty.

State aid and its control will be of great importance for the new Member States in terms of regional development, on the one hand, and in view of the adjustment process of their industry, on the other hand. This process will not have been completed totally at the time of accession.

In recognition of special requirements of less developed regions, former accession treaties (such as the treaty with Spain and Portugal) included a protocol stating that the requirements of economic and social development in these regions would be taken into account when deciding on the admissibility of state aid. The EU Commission is anyhow *empowered* to base its decisions on such considerations when deciding on state aid since it has a broad discretionary power under Art. 85 EC-Treaty. The protocols mentioned go one step further, they *oblige* the Commission to do so.

As has been mentioned, the restructuring of the economy in the countries of Central and Eastern Europe poses problems which are different from the 'normal' tasks of regional development. If a comparable protocol were attached to the new accession treaties it would have to refer to this special situation. Such a protocol has been requested by some of the applicant states. In view of the practice in former accessions, this request seems to be justified. It is difficult to see how the EU could refuse it in this case.

Whether a protocol of this kind is included into the accession treaty or not, the Commission will have to take the special situation of the Central and East European countries into account when deciding on state aid proposed by these countries. It is in the interest of the EU that these countries can continue on the path of economic development and that this growth is not disrupted.

This is not to say that these states can or should receive a *carte blanche* for handing out state aid. The control of the Commission will have to be maintained. And the aim cannot be to support and encourage unsuccessful enterprises that would, under normal circumstances, stop operating and fall into bankruptcy. The assessment in individual cases will be difficult, and there is a narrow border line between being too 'lenient' and the unjustified application of the 'normal' EC rules to cases which are of a different nature.

The Lisbon European Council concluded that in the area of state aid the emphasis should be shifted from the support of individual companies to more general programmes of Community interest, such as employment, regional development, environment, and education and research. This approach is to be supported, and the accession states are well advised (also for economic and financial reasons) to orient their state aid policy along these lines.

It should not be overlooked, however, that economies in transition also require ad-hoc-aid to individual enterprises in difficulties or in the process of restructuring which normally should not be granted. The EU policy on state aid will have to take this into account for a certain period until the process of readjustment of these economies is completed.

Environmental Policy

The environmental policy of the EU covers many areas. There are directives of a more general nature and directives for specific sectors. Some directives are product-related, others are applicable to installations or their operation. There are directives concerning industry and trade or agriculture, whereas others lay down more general standards or concern specifically nature protection.[16]

In general, accession will have no major impact on the EU's environmental policy and its further development. The inclusion of new Member States and the application of the acquis by them will have a very beneficial effect on the environment in at least some of the present Member States as well. It will be necessary, however, to grant long transitional periods to the new Member States and also to differentiate, at least as to the date of its application, as far as new environmental measures are concerned which the EU might take in the future.

Adoption of the Acquis and Accession

It is not surprising that the acceptance of the acquis poses a number of problems for the applicant states. They have requested numerous transitional periods of varying length.

The application of most of the directives under this heading can be costly and requires investments of billions of Euro. In addition, its implementation requires a highly skilled administration. The new Member States will not be able to conform to the environmental acquis in all its aspects at the time of accession – if this takes place at any foreseeable date, as is to be hoped.

These problems should not be used as an argument against accession in the near future. Most of the negative consequences of not conforming to the acquis at an early date are independent of accession and would be felt in the EU and by its enterprises to an equal extent also without accession. With a free trade area already in place, competitive disadvantages for EU business would be felt whether there is accession or not.

An early accession would most likely accelerate the adoption of the acquis and thus have positive results also for the EU. The increase of funds made available by the EU after accession will support this effect.

Costs and Benefits

In view of the considerable cost of introducing the environmental protection measures of the EU, it is important to examine in each case whether the fulfilment of the EU standards is the optimal resource allocation in the new Member States.

A cost-benefit-analysis (Art. 174 para. 3 EC-Treaty) must include alternative allocations of the limited resources available. This is of importance also in relation to other parts of the acquis which the new Members will have to adopt and implement, for instance in the area of transport policy (TEN – see below). The EU recognises this in its Common Position on the Polish Paper for the chapter on the environment (Conference on Accession to the European Union-Poland-EU Common Position, doc. CONF-PL-55/99).

In a number of cases, a relatively high standard of environmental protection can be achieved at a still acceptable cost. The complete adoption and implementation of the acquis in all areas and for all installations, on the other hand, could incur expenses which might be – under the given circumstances – out of proportion to the additional improvement which can be achieved.

If at all possible, a priority scale should be established for the environmental measures themselves. The following list (indicating decreasing importance) gives an example of what could be done, taking in account the specific interests of the present EU and its Member States (competition, trans border pollution) and not primarily the interest in environmental protection per se:

- Product-related standards, because they effect the functioning of the internal market.
- Horizontal provisions, such as the Directive 85/337/EEC on Environmental Impact Assessment because they have a general impact on a number of public and private projects.
- Production-related standards, because of their effect on competition in the internal market.
- Standards relating to pollution with cross-border-effects, because of this effect.
- Other rules, such as the directives on drinking water or bathing water, which are primarily of national significance.
- Nature protection.

Whatever the priority according to such a list, *new* installations should conform to EU standards right from the start. The accession countries would be well advised to introduce this rule as early as possible; in many cases they have done so already.

The optimal allocation of the resources in the new Member States (and, possibly, also in some of the existing Member States) might require a differentiated approach also in the future. It cannot be in the EU's interest if the introduction of ever higher standards – as justified as they might be – impedes the economic development of a country. New EU measures which are adopted after accession might also have to include transitional periods or other special provisions for the application of the new rules by the new Member States.[17]

Cases in point are the drinking water directives and the directive on bathing water. Starting with the latter one first: If the principle of subsidiarity (Art. 5, EC-Treaty) were taken seriously, the EU should not have a bathing water directive at all. The objectives of that directive can be achieved sufficiently by the Member States themselves, except for the limited number of cases of cross-border relevance. Because of this, its implementation by the new Member States cannot be a priority.

The case of the drinking water directive is somewhat different: Drinking water per se is of local concern only and could be handled by the Member States. Its quality is of EU concern in areas of relevance for the internal market: with good reason, the EU sets standards for water used in the production of goods which enter the internal market. That should be the focus of any EU rules on drinking water.

The drinking water directive, in addition, is an example of measures where achieving the required highest standard is relatively costly whereas the achievement of the – still acceptable – first stage could be done at relatively lower cost.

Environment and Competition

It is evident that a number of the environmental measures of the EU have an impact on competition. All product-related measures can be left aside since their adoption by new Member States will be necessary and, in principle, cannot be delayed because of the internal market.

Environmental requirements related to the construction or the operation of plants are often an important cost factor. If plants in certain Member States or certain regions are exempt from applying the common standards, this is a financial advantage which is of relevance for their competitiveness.

As in other areas of policy, the competitiveness of any business is determined by a whole range of factors. None of these factors can be looked at in isolation. For instance, it may be assumed that productivity is low for a number of reasons in many of the plants in the accession countries. They might find it hard to compete in the internal market even if the wage levels remain relatively low for some time to come. They might 'need' the relative and 'artificial' advantage of not complying to environmental standards for some time in order to be competitive at all. Enforcing EU environmental standards on existing smaller businesses might jeopardise their existence, if no sufficient time for adjustment is granted. It is not in the EU's interest to endanger the economic recovery and development in these countries.

Employment Policy and Social Policy

Employment Policy

A chapter on Employment Policy was introduced into the EC-Treaty by the Treaty of Amsterdam (Art. 125 ff). Essentially, this chapter provides for the co-ordination of the employment policy of the Member States: 'Member States and the Community (...) shall work towards developing a coordinated strategy for employment' (Art. 125, para. 1).

This chapter establishes certain objectives to be pursued by the Community and by Member States. It does not prescribe how these aims are to be realised, at least not in detail. The Treaty provisions do not preclude that special situations in new Member States (as in different old ones) are taken into account when formulating the national employment policy and when coordinating employment policies on the EU level.

The Council agreed on 'Guidelines for Member States' Employment Policies for the Year 2000'. They deal in four chapters with improving employability; developing entrepreneurship; encouraging adaptability of business and their employees; and strengthening equal opportunities for women and men.

The guidelines state expressly that 'the implementation ... may vary according to their nature, the parties to whom they are addressed and the different situations in the Member States' (Council Decision 2000/228/EC, 2000). There is, basically, no reason why new Member States from Central and Eastern Europe should not be able to subscribe to such guidelines and to implement them.

In view of the development of their economies, the restructuring process and their social systems, specific elements might be added once these countries are included into the decision making process as (new) Member States. This, however, will not have a major impact on the EU employment policy as it has been developed so far.

The same holds true, in principle, for the Lisbon European Council conclusions on employment policy. Some accession countries might find it difficult to meet one or the other of the targets which the European Council set for measures in the field of education and training. Problems in these areas could be accommodated by introducing transitional periods without changing the substance of the EU policy.

The Lisbon European Council and the guidelines refer to the 'social partners', either by including them into the process of developing and executing the guidelines, or by leaving certain tasks up to them. In some of the accession countries the system of 'social partners' has not been developed as yet to the same extent as in most countries of the EU-15.

The role of the different social partners, and of 'social partnership' itself varies also among the present Member States. It should not present a major problem, therefore, to accommodate different models/practices from some of the new Member States within the framework of the EU's employment policy.

Social Policy

I do not propose to discuss the question of free movement of labour which has been raised in connection with the accession of the Central and East European countries. This is essentially a question of transitional periods, i.e. the impact on EU policies is temporary. There are no proposals – and it would probably not even be possible – to limit the free movement of labour permanently in relation to the new Member States.

The social policy of the EU encompasses a number of directives harmonising national law which cover especially the harmonisation of labour law and the protection of health and safety.

In connection with the accession of the Central and Eastern European states, two elements of this legislations are of particular importance and might pose problems: Competitors from the EU-15 will raise the question of fair competition, and they will be interested in the full application of these measures also by enterprises in the CEE states. The enterprises in the CEE states, on the other hand, might find it difficult to apply strict social standards at a time when their restructuring process still has not been completed.

The discussion on the development of the EU's social policy has always included the aspect of 'fair competition'. The (misleading) term 'social dumping', which was even introduced into the WTO discussions, testifies to that.

The feeling prevails that within a functioning internal market social rules also need to be harmonised. Often enough, however, it is overlooked in this context that less developed economies would have enormous problems in EU-wide or world-wide competition if they could not use lower social standards than the more prosperous countries to offset other drawbacks.

The level of social standards is directly related to the level of prosperity in the different countries or regions. Imposing high social standards on an economy which cannot cope with them might have a very detrimental effect on the further economic – and thus also social – development of that country or region.

In Germany, it is generally assumed that salaries in the new *Laender* rose too fast and ahead of productivity gains. This might have been inevitable for political reasons. Still, this experience should give one pause to think. I am not pleading for low social standards. I am far from doing that. But diversity in social standards is not necessarily an infringement of fair competition. To the extent that the accession countries request transitional periods, such problems can be accommodated within that framework without having a lasting impact on the EU's social policy.

In assessing a possible impact of accession beyond such transitional periods, it should be kept in mind that Art. 137, para. 2 provides that such measures should establish only '*minimum* requirements for *gradual* implementation' (my emphasis). This provision does not aim at fixing only requirements which also the least productive economy or enterprise can fulfill. The main reason for this provision consists in allowing Member States to go even beyond the level which has been established by the EU directives. Still, in general this provision will ensure that the EU does not impose the highest standards on all the Member States without distinction. The 'gradual' implementation may be considered as another element of built-in elasticity. Moreover, the same article of the EC-Treaty states that the Council should have 'regard to the conditions (...) obtaining in each of the Member States'.

Some of the secondary measures taken by the EU include clauses that allow Member States to opt for different solutions or derogations concerning certain provisions, as for instance the directive on working time (93/104/EC). However, there is no general reference to adjustments which might be necessary because the economy or enterprises of a certain Member State cannot support the standards laid down in the directive.

It might very well be that after accession the new Member States will argue in certain cases for introducing lower standards than would have been agreed on otherwise, or that they will ask for delays in the application of the directive. If such problems should arise, they would be of a transient nature. Assuming that these states will narrow or even bridge the gap between their economic development and that of the present Member

States, the need for special rules applicable in the new Member States will diminish.

If the EU takes diversity into account to a greater degree than it has till now, it should be able to pursue its social policy as already established and as envisaged for the future.

Transport Policy

The accession of the Central and East European countries as well as Malta and Cyprus is a challenge also for the transport policy of the EU. I propose to leave Malta and Cyprus aside here since their inclusion in the EU transport policy is largely limited to sea transport. I further limit my discussion to transport infrastructure since the impact on EU policies will be most pronounced in this area of transport policy.

The ten continental accession states cover an area of slightly above one million square km. This represents almost a third of the area of the present EU-15. The transport infrastructure in most of these countries is still underdeveloped, if measured by standards of the EU-15. This concerns practically all modes of transport and is especially pronounced for the road and rail network.

In assessing the need for improving the transport infrastructure in these countries, it is necessary to take into account that most of these countries are also important transit territories for traffic either to Russia, Belarus and the Ukraine, or to Turkey. Also this traffic is likely to grow considerably in the future.

Economic development of a region depends to a great extent on the availability of an adequate infrastructure. The Community guidelines for the development of the trans-European transport network (TEN-T) of 23 July 1996 (Decision No. 1692/96) refers explicitly to the 'smooth functioning of the internal market and the strengthening of economic and social cohesion'. According to these guidelines the TEN-T 'shall be established gradually by 2010 by integrating land, sea and air transport infrastructure networks throughout the Community'.

Spain may be cited as an example of what can and needs to be done in that country with considerable EU support. In Germany, the new *Laender* are an example which is possibly more comparable to the Central and East European Countries, even if the reconstruction of infrastructure there was mainly financed through national funds.

Developing the transport infrastructure must, therefore, be considered as a priority. The EU has responded to this need by making funds available for transport infrastructure measures through ISPA (Council Regulation No. 1267/1999). This covers the time until accession.

As in the cohesion policy in general, the EU will have to place a special emphasis on the needs of the CEE countries. It would be well advised to increase its support for such measures even before accession and beyond the amounts which are channeled through ISPA – provided the countries show a sufficient absorption capacity.

The EU should decide to employ toward this end some of the funds which were agreed on at the European Council in Berlin for the period after accession (beginning in 2002). It would also be in the interest of the EU to increase the funds devoted to infrastructure gradually before accession, assuming that accession will not have become effective in 2002. It is doubtful whether a sudden increase after accession at a later date would allow for an economical expenditure of the funds.

The impact of accession on the transport infrastructure policy of the EU will probably not consist in *changing* this policy. I do not see a major need for that. But the emphasis of this policy will have to shift to the new territories of the EU and their particular needs which are more pressing than in the old Member States.

In extending the TEN to the new CEE Member States, the ten Pan-European Transport Corridors may be considered to form the backbone of the network in these countries. These corridors are the extension of the traffic arteries of the EU into Central and Eastern Europe. They include such connections as Helsinki – Warsaw, Berlin – Warsaw – Moscow, Dresden – Prague – Budapest – Sofia – Istanbul, Venice – Ljubljana – Budapest – Lwow, and Salzburg – Ljubljana – Zagreb – Thessaloniki.

A Transport Infrastructure Needs Assessment (TINA) for the accession countries has been completed. This was largely based on the same criteria as those for the TEN-T Guidelines. Apart from the CEE countries, this work was also based on contacts with Russia, Belarus and the Ukraine.

The TINA countries estimated the necessary investment (reconstruction, upgrading, construction) at more than € 90 billion through to 2015, out of which € 44.3 billion are for the road network and € 37.1 billion for the railway network.

The ISPA funding for transport infrastructure will be about 520 million €/year through 2006. Assuming there will be a sufficient absorption capacity and that the required additional financing can be secured, there is certainly a need for higher resources. After accession, there will be more resources available, either from the structural funds, the cohesion fund, and/or the TEN programme.

The support for infrastructure from the structural funds and from the cohesion fund is not limited to the TEN. As important as the TEN are, especially in fully integrating these countries into the internal market, they can gain their full effectiveness for the EU as a whole and especially for these countries only if they are based on a network of transport infrastructure covering the whole country.

Conclusions

The following conclusions can be drawn:

- Policy making in the greater EU will have to take diversity more strongly into account.
- This does not have to be seen as endangering integration.
- Proposals for a two-tier-Europe or a 'core-group' within the EU should not be supported.
- For some policy areas, taking accession into account offers the possibility of making *better* policy.
- The smooth functioning of the internal market must be safeguarded.
- The budget allocations will have to be adjusted and shifted more strongly toward the new Member States, without raising the own resources ceiling.
- The structural measures will have to include fully the new Member States, with greater emphasis to be placed on the Member States/regions at the lower end of the scale of economic development.
- The reform of the Common Agricultural Policy will have to be continued with greater zeal, and it will have to take accession into account.
- The reform of the CAP in all its aspects should be pursued at the mid-term review (2002/2003) so that it can be completed before accession.
- The application of the rules on state aid must take into account the special situation of economies in transition.

The impact of accession on EU policies is no reason not to admit new Member States or to delay accession.

Notes

1   This does not apply in quite the same way to Slovenia, and also not to Cyprus and Malta.

2   This paper does not go into the institutional adjustments which are the topic of the present Intergovernmental Conference. These reforms may be counted among the impacts of accession, but they are not impacts on EU policies.

3   This does not mean, however, that every measure based on Art. 95 EC-Treaty is of the same significance. Does the prohibition of cigarette advertising need to apply at the regional and local levels, for instance?

4   Item 2 of the Protocol on Subsidiarity, which states that the acquis remains fully safeguarded, must also be understood in this way, i.e. as referring to the more important rules. The motive for this provision was to prevent the so-called re-nationalisation of matters that have already been regulated by the EU.

5   The EU Commission's communication on the Agenda 2000 mentions about € 11 billion per year until 2005.

6   A recent calculation predicts for the 5 CEE countries of this group for the year 2006 a total GDP of € 320.955 million; four per cent would amount to € 12.838 million. See, *Gutachten der Friedrich-Ebert-Stiftung, Die Osterweiterung der Europaeischen Union: Konsequenzen fuer Wohlstand und Beschaeftigung in Europa*, March 2000.

7   In its communication on the Agenda 2000 the EU Commission expressed its view that enlargement could be financed within the present own resources ceiling of 1,27 per cent of GNP, depending on certain conditions.

8   The European Council laid down in greater detail the rules for the Structural Funds and the Cohesion Fund (because they affect the distribution of resources between the Member States) and some rules for the Agricultural policy.

9   Cyprus and Malta are not included in this survey.

10   There is a slight difference between the wording of the two provisions. I propose to disregard this discrepancy in the context of this chapter.

11   It is of interest to note that the future development of the Common Agricultural Policy might influence the GDP/head of a given region: To the extent that high price support is replaced by compensatory payments to the farmers, the GDP of that region might go down because the compensatory payments are not counted as 'domestic product', even if the income of the farmers remains basically unchanged.

12   To provide an example: In the case of Ireland, the transitional support is higher than the support for the Irish regions remaining under Objective 1 (€ 1.773 to 1.315 million). There is no reason why such a considerable transitional support should be necessary under the heading of 'cohesion'. This is all the more so since, again in the case of Ireland, the criteria for the objective-1-regions had already been surpassed for some time (the GDP/head in Ireland came close to the EU average which means that it was well above the EU average in the more developed regions of Ireland!) –

and Ireland must have been well aware that no more support under this heading could be given to those regions after 1999.

13   See Draft Common Position for Poland on Negotiation Chapter 7, Agriculture, draft 29/03/2000, p.4.

14   See EU Commission, Draft Common Position on Poland, Negotiation Chapter 7, Agriculture, Draft 29/03/2000.

15   One has furthermore to keep in mind that there are interrelations between some of the provisions of this regulation with the regulation of the milk market and its quota system. If the latter is to be reformed, this will have consequences for the regulation on beef and veal.

16   The Commission Staff Working Document 'Guide to the Approximation of European Union Environmental Legislation' of July 1997 provides a good overview for the associated countries.

17   If that is done, the EU Commission's fear does not seem to be justified that accession might impede the further development of the EU's environmental policy (Communication on the Agenda 2000).

References

EU Commission (1997), *Guide to the Approximation of European Union Environmental Legislation*, Staff Working Document of July 1997, P 02.18.091 GUID4.DOC.

EU Commission (1998), *Communication on the Agenda 2000*, 16 July, 1997.

EU Commission (1999), *Report on the Execution of the Budget of the Commission of 31 December 1999* (preliminary version), SEC (2000) 340.

EU Court of Auditors (2000), *Annual Report on the 1998 Budget*, Preliminary Edition,

Eurostat (2000), *News Release* No. 48/2000 of 18 April, 2000.

Friedrich-Ebert-Stiftung (2000), *Gutachten: Die Osterweiterung der Europaeischen Union: Konsequenzen fuer Wohlstand und Beschaeftigung in Europa*.

4 Enlarging EU Environmental Policy: The Challenges of Flexibility and Integration

INGMAR VON HOMEYER

Introduction

When one looks at the larger political context of contemporary European transport policy, two developments appear to be particularly important: First, there is a growing awareness that environmental concerns must be more systematically integrated into transport related decision-making at the European level. Efforts are already under way in this respect in the framework of the so-called 'Cardiff Process'. Second, the end of the Cold War and the upcoming enlargement of the European Union (EU)[1] to include ten Central and Eastern European Countries (CEECs)[2] have fundamentally altered basic parameters of European transport policy. Against this background, and based on a detailed analysis of the implications of enlargement for European environmental policy, this chapter argues that in addition to posing a number of more general environmental policy challenges, enlargement further increases the need to systematically integrate environmental concerns into transport policy.

Eastern enlargement of the EU has often been described as a major challenge for European environmental policy (cf. EC, 1998a; Carius *et al*, 2000). In its Agenda 2000 the European Commission concluded that 'none of the [Central and Eastern European] Accession countries can be expected to comply fully with the [environmental] *acquis* in the near future, given their present environmental problems and the need for massive investments' (EC, 1997, p.67). In a similar vein, Jordan points out that the 'impending enlargement of the EU to encompass countries from the former Eastern Bloc constitutes a far stiffer challenge to the internal cohesiveness and high aspirations of the environmental *acquis* than anything that has gone before, including tardy implementation and fitful policy integration' (Jordan, 1999, p.15).

99

These views beg the question why the failure of the Accession Countries to fully adopt European environmental legislation at the date of accession (assuming that accession takes place in 'the near future') should pose a major challenge to European environmental policy. After all, the Accession Countries may still complete the process of implementation of EU environmental legislation after accession.

To answer this question it may be necessary to take a broader look at environmental policy-making in the EU and to study challenges from a perspective which takes additional, contextual factors into account. These factors will have an important effect on the impact on EU environmental policy-making of enlargement and the Accession Countries' likely failure to achieve full and timely compliance with European environmental legislation.

More generally, it is only possible to analyse the implications of enlargement for EU environmental policy if one has an idea of the central elements which characterise European policy-making in this field. This chapter therefore begins with a detailed analysis of three basic characteristics of contemporary EU environmental policy-making. In a second step, I then look at the implications of enlargement for these characteristics.

It has almost become a common place to note that in the first years of existence 'the EU had no environmental policy, no environmental bureaucracy, and no environmental laws. Twenty five years later, the EU has some of the most progressive environmental policies of any state in the world although it is not a state. EU environmental policy has now achieved a coherence and legal structure that is unique among international organisations' (Jordan, 1999, p.1).[3]

As illustrated in more detail below, the first factor which has a crucial impact on the dynamics of EU environmental policy-making is the so called 'leader-laggard dynamic', primarily because this mechanism may account for the 'tremendous expansion of environmental policies' (Zito, 1999, p.19). At the most fundamental level, the leader-laggard dynamic is based on regulatory competition between environmentally more progressive 'leader' Member States and a second group of 'laggard' Member States which tend to prefer a relatively low level of environmental protection.

In addition to the leader-laggard dynamic and the resulting proliferation of EU environmental legislation, European environmental policy is characterised by serious implementation problems.[4] For example,

Richardson notes that 'the environmental field is a spectacular example [of unsatisfactory implementation of EU policies], with almost every Member State being brought before the European Court of Justice for failure to properly implement certain aspects of EC environmental law' (Richardson, 1996, p.284).[5] Although other areas of EU policy-making are plagued by similar implementation problems, the deficit appears to be particularly serious in the case of environmental legislation (cf. Jordan, 1999a, p.83; Collins and Earnshaw, 1992).

As we will see, to a large extent the implementation deficit reflects the success of European environmental policy at the legislative level. Those Member States which failed to influence EU environmental policy-making as either leaders or laggards – and for which implementation may be associated with relatively high financial costs or administrative problems – are particularly prone to delayed or incomplete implementation. Against this background the increasing use of self-regulatory and more flexible instruments in EU environmental policy frequently reflects an effort by the Commission to reduce the implementation deficit (cf. Knill and Lenschow, 1999, pp.592-593; Golub, 1998a).

However, the implementation deficit is not the only problem which casts doubt on the record of EU environmental policy so far. It is becoming increasingly evident that despite the rapid proliferation of EU environmental legislation the overall state of the environment in the Member States has not significantly improved. Although there has been some notable success in combating certain environmental problems, for instance traditional air and water pollution, the deterioration and destruction of eco-systems and the resultant loss of biodiversity has, if anything, only been slowed down (EEA, 1999, pp.23-37; Buck, Kraemer and Wilkinson, 1999, p.12).

The third fundamental characteristic of contemporary EU environmental policy-making relates to the political efforts by Member States and the Commission to reform EU environmental policy which may be interpreted as a reaction to the fact that EU environmental policy has so far failed to halt overall environmental degradation. As illustrated further below, the integration of environmental concerns into sectoral policies is probably the most important and promising feature of ongoing efforts to increase the effectiveness of EU environmental policy.

The three characteristics only provide a somewhat sketchy picture of the state of contemporary EU environmental policy-making. Nevertheless, they illustrate aspects which have been very significant in the past. More importantly, I argue that these three factors can be expected to gain additional importance as a result of enlargement.

The next section sets out the basic analytical framework for this argument. It focuses, in particular, on the leader-laggard dynamic as a crucial driving force of EU environmental policy-making. The third section discusses the extent, the causes, and the implications of the implementation deficit. Section four highlights the need to integrate environmental concerns into sectoral policies. Against this background, section five focuses on the implications of enlargement for the three fundamental characteristics of European environmental policy and the increasing need to integrate environmental concerns into sectoral policies, in particular transport policy.

Regulatory Competition

In a seminal article from 1988 (German original 1985) on what he called the 'joint decision trap' Scharpf provided an explanation for the difficulties encountered by actors attempting to reform inefficient EU policies, in particular the Common Agricultural Policy (CAP) (Scharpf, 1988). According to Scharpf, unanimous voting rules in the Council of Ministers and the institutional self-interests of Member State governments account for the failure of the EU to adopt new, or reform existing policies. Unanimous voting rules make it difficult to adopt European legislation which exceeds the lowest common denominator of Member State interests. At the same time the institutional self-interests of Member State governments in retaining their veto rights prevent a reform of decision-making in the Council. As a result, stagnation and inefficient policies dominate EU policy-making.

Scharpf's account of decision-making obviously cannot explain the dynamic development of a comprehensive and advanced body of European environmental legislation.[6] Although several explanations have been proposed for the development of EU environmental policy,[7] competitive interaction between environmental leader and laggard Member States is generally considered to be one of the most significant factors.[8] As early as the beginning of the 1980s analysts classified the Member States according to their generalised support for ambitious EU environmental policies and proffered tentative explanations of EU environmental policy-making in terms of the relative influence of leaders and laggards among the Member States.[9]

It is possible to distinguish ecological, economic, administrative, and political influences on the interests of Member States in EU environmental policy formation.[10]

Ecological Influences

Ecological determinants of Member State interests reflect various factors influencing the state of the environment. Geographical conditions or basic characteristics of the economic structure of a Member State tend to be particularly relevant. For example, Germany, and other Member States in the north-east of the EU, generally tend to prefer uniform European emission standards over environmental quality standards because they suffer from transboundary pollution as a result of geographical and climatic factors. By contrast, the western Member States often prefer quality standards because they face considerably less transboundary pollution due to more favourable geographical and climatic conditions (cf. Golub, 1996, pp.707-709). Uniform emission standards would have the effect of forcing these countries to comply with a relatively high level of environmental protection which responds to the needs of the more heavily polluted north-eastern Member States. In contrast, the use of environmental quality standards allows such countries to fully benefit from their geographical and climatic advantages.

Ecological influences on Member State interests also result from differing economic structures. For example, due to heavy industrialisation many northern Member States have a stronger interest in combating industrial pollution than the less industrialised southern Member States (cf. Holzinger, 1993, p.77).

Political Influences

Environmental problems do not always lead to political action. Often, there is a considerable mismatch between the degree of environmental pollution and the level of political attention which the pollution receives. In fact, political action frequently occurs in situations in which environmental pollution is relatively low or decreasing, while more severe environmental problems may receive less attention (cf. Prittwitz, 1990). However, political factors may also be important in overcoming economic and other interests which mitigate against measures to protect the environment (Scharpf, 1998, pp.129-132). A range of political factors, such as the respective party system, the presence of a strong environmental movement, influential environmental interest groups or parties, and wide-spread environmental consciousness among the general population all have an influence on whether and how environmental problems are translated into Member State interests. The German position at the EU level on the planned introduction of significantly stricter emission limits for large combustion plants and cars in the 1980s exemplifies the relevance of domestic politics. While such measures had previously been strictly opposed by the German government, the first electoral successes of the Green party fundamentally changed the position of the government on this issue (Boehmer-Christiansen and Skea, 1991; Andersen, 1997, p.213; Sbragia, 1996, pp.248-249).

Economic Influences

Economic considerations play a major role in the formation of Member State interests in EU environmental policy. Scharpf supplemented his argument regarding the joint decision trap to take account of certain economic factors which may explain why EU environmental standards are relatively strict. He distinguishes between environmental 'product' and 'process' standards.[11] Regarding product standards, it is unlikely that European regulations reflect the lowest common denominator of Member State interests because the economic advantages of a single standard are an incentive for environmental laggard Member States to enter into a compromise with environmental leaders, in particular in cases where the leaders control export markets important for the laggards (cf. Scharpf, 1994, pp.233-234).[12] However, regarding process standards, there are usually only few economic incentives for laggard Member States to adjust their standards upwards because it is much more difficult for leader countries to keep products which were produced subject to low production standards from entering their markets.[13] On the contrary, if lower standards have a significant effect on production costs, economic competition is likely to result in a 'race to the bottom' where Member States compete for the lowest production standards in order to promote their industries (cf. Scharpf, 1998, pp.131-132; Scharpf, 1994, pp.234-235).

Similarly, the debate about emission vs. quality standards is not only motivated by different levels of transboundary pollution, which may be attributed to geographical and climatic factors, but also by economic considerations. For example, British opposition to emission standards is largely based on the view that compliance with relatively high emission standards means that British industry is burdened with costs which are not justified given the level of local pollution in the UK (cf. Golub, 1996, pp. 707-708).[14] Conversely, Member States which have adopted, or plan to adopt, relatively high emission standards have an economic interest in other countries adopting the same standards in order to prevent their own industries from becoming less economically competitive (cf. Weizsäcker, 1990, p. 44; Scharpf, 1997, p.20).

Finally, different levels of economic development also constitute an important influence on Member State interests in EU environmental policy. Although a high level of environmental protection may in the long term frequently be associated more with economic advantages than with higher costs (OTA, 1994), the financial requirements of implementation can be substantial in the short and medium term. As a result, less economically developed countries tend to be laggard Member States which oppose the adoption of strict environmental standards at the EU level (cf. Holzinger, 1993, p.77).

Scharpf points out that 'if differences in the level of economic development, and hence, industrial productivity, would create differences in the ability of national economies to absorb the costs of regulation [...] less developed countries must resist uniform regulations defining levels of regulation that correspond to the willingness and ability to pay of citizen in more highly developed [and, generally, more highly polluted] countries' (Scharpf, 1996, p.15; see also Sbragia, 1996, p.249). Put differently, assuming that pollution levels are similar in developed and less developed Member States, if the short and medium term costs of environmental protection absorb a higher share of available resources in the less developed Member States, then these countries will have an economic incentive to oppose the adoption of higher EU standards. In fact, the poorer Member States, such as Portugal, Spain, Greece and Ireland, are usually counted as environmental laggard countries.

Administrative Influences

National administrative traditions may influence Member States' interests in EU environmental policy. Regarding environmental policy, German and British interests have so far been particularly strongly associated with the promotion and diffusion of national administrative traditions. Administrative arrangements in these two countries may be characterised as 'diametrically opposed' with regard to issues such as the use of emission as opposed to quality standards, hierarchical substantive regulation as opposed to procedural self-regulation or formal as opposed to informal interest intermediation (Knill and Lenschow, 1997, pp.1,5). These differences contributed to the development of competition between these two countries to influence EU environmental policy-making.

More specifically, in the early 1980s successive German governments tried to 'Europeanise' German environmental policy by means of political initiatives at the EU level and in other international organisations (cf. Boehmer-Christiansen and Skea, 1991, p.193; Sbragia, 1996, p.240). In addition to an economic interest in avoiding competitive disadvantages for German industry as a result of the planned introduction of stricter environmental legislation in Germany, this strategy aimed at minimising the administrative adjustment costs which were expected to follow from future European regulations (Heritier, Knill and Mingers, 1996, pp.175-176; see also Zito, 1999, p.26).

The 'greening' of the German government in the early 1980s was followed by a similar development in the UK in the late 1980s (cf. Golub, 1996, p.711). Britain copied the German strategy and managed to export its regulatory approach via the EU to the other Member States. Once more, the aim was to reduce the economic burden for domestic industry and to minimise administrative adjustment costs (Heritier, 1995, p.294).

Although a large number of ecological, political, economic and administrative factors influence the formation of Member State interests in EU environmental policy, it has nevertheless been possible to identify several environmental pace-setters. The six environmental leader Member States Austria, Denmark, Finland, Germany, the Netherlands, and Sweden frequently support the adoption of relatively strict environmental legislation at the EU level. With respect to these countries a number of factors which tend to support a Member State's interest in high level EU environmental legislation converge. These factors include a high GDP, advanced industrialisation, high affectedness by transboundary and/or domestically caused pollution, wide-spread environmental consciousness, and high domestic political relevance of environmental policy (cf. Holzinger, 1993, pp.78-79). Conversely, factors supporting a Member State's opposition to strict EU environmental legislation – e.g. a lower GDP, lower industrialisation and pollution levels etc. – tend to converge in the case of the environmental laggard Member States Greece, Ireland, Portugal, and Spain.

Interestingly, the U.K., which until recently also belonged to the group of laggards, differs from the rest of the group with regard to a number of factors. Britain's initial reluctance to support high EU environmental standards may be attributed to a large extent to the special geographical situation of the country, to particularly severe economic difficulties in the 1980s, and to a regulatory approach which differs sharply from the German administrative tradition. However, the same factors may also explain why the U.K. has gradually adopted a more constructive attitude since the late 1980s: Not only has the economic situation in the UK drastically improved but, perhaps more importantly, the present administrative requirements of EU environmental policy are much more similar to traditional British administrative practice than they used to be in the 1980s.[15]

The remaining Member States – e.g. Belgium, France, Italy and Luxembourg – either choose to support environmental leader or laggard Member States on a case-by-case basis or opt for a medium level of EU environmental regulation (cf. Holzinger, 1993, p.78).

The fact that Member States can be classified as environmental leaders and laggards does not in itself imply that the joint decision trap can be easily overcome, in particular because coalitions between leader or laggard Member States are not permanent but have to be negotiated on an *ad hoc* basis (Liefferink and Andersen, 1998, p.264). The various factors which account for Member State interests in EU environmental policy-making merely suggest that the institutional self-interest in retaining decision-making powers at the national level may be offset by a host of other interests. In addition to the interests arising out of ecological, political, economic and administrative factors, institutional arrangements at the European level contribute to overcoming the decision-making constraints which characterise the joint decision trap, in particular those caused by unanimity voting.[16]

Institutional Factors Influencing Regulatory Competition

Highly complex institutions, such as the multi-level decision-making system of the EU (Marks, 1993; Hooghe, 1996), not only provide for veto points which enable actors to block decisions (Immergut, 1992, pp.226-244) but, perhaps more importantly, also offer multiple channels for sufficiently resourceful and motivated actors to push their preferred solution through the decision-making system (cf. Peters and Guy, 1992, p.118; Mazey and Richardson, 1993, p.112). As pointed out by Heritier (1998, p.4), the deadlock-prone formal decision-making system of the EU is particularly likely to develop more or less informal 'second-order' institutions which allow actors to overcome deadlock. Frequently, the ability of environmental leader Member States to influence EU environmental policy-making depends on the availability of such channels.

The introduction of qualified majority voting (QMV) to the Council of Ministers by the Single European Act (1986), and its subsequent expansion in the Maastricht Treaty (1992), and the Amsterdam Treaty (1997) to cover more and more areas of environmental decision-making is generally seen as a particularly important factor in overcoming resistance by environmental laggard Member States to efforts to introduce high EU environmental standards. However, even under QMV it is relatively easy for environmental laggard Member States to block legislative proposals by leader countries (cf. Andersen, 1997, p.218). Moreover, if leader Member States are unable to form a blocking minority, they may even be forced to accept a legislative outcome which reflects the lowest-common denominator of the interests of a coalition of laggard Member States and some of the neutral countries (Holzinger, 1994, pp.466-468; Golub, 1996a).

In addition to QMV, a number of other institutional arrangements at the EU level further increase the influence of leader Member States on EU environmental policy-making. The Commission's agenda setting power is one of the most important factors (cf. Pollack, 1999). Due to the fact that the Commission has relatively few staff and lacks in-house expertise, it is heavily dependent on Member State experts in its numerous advisory committees (Heritier *et al.*, 1996, pp.152-153). The Commission also employs temporary national experts, who are seconded from their home administrations. Member States frequently use this opportunity to 'parachute' experts into strategic positions in the Commission to advance their domestic regulatory approach and standards (cf. Liefferink and Andersen, 1998).[17] There are also channels for Member States to submit written proposals for EU environmental legislation to the Commission, for example through the Environmental Policy Review Group (EPRG) in which high level Member State officials are represented (Liefferink and Andersen, 1998, pp.264-266).

These channels of access to the Commission would only be of relatively minor importance for environmental leader Member States if they were not coupled with a specific set of institutionally shaped preferences of the Commission and its services. First, as the 'motor of integration' the Commission has an institutional self-interest in expanding its regulatory competencies (Majone, 1994, pp.31-33). All things being equal, the Commission is therefore more likely to accept policy proposals by Member States if these proposals offer opportunities for an expansion of its tasks. Such proposals tend to come from environmental leader Member States rather than from laggard countries.

These dynamics can be most clearly observed in the sustained efforts by laggard Member States to block the adoption of EU environmental legislation. It should be mentioned, however, that in some of the more extreme cases leader Member States may also oppose EU legislation. For example, in the early 1990s several Member States, including the UK and Denmark, invoked the subsidiarity principle to call for a repatriation of environmental competencies (cf. Collier, 1996, p.12). In addition, some leader states emphasise the 'environmental guarantee' of Article 95 (4) EC (cf. Liefferink and Andersen, 1998, pp.258-259). Under certain conditions this provision gives a Member State the right to exceed EU environmental standards (Albin and Bär, 1999). However, although upward deviance from EU standards by environmental leader Member States tends to undermine EU competencies in the short term, it may lead to renewed calls for the adoption of harmonised European regulations in the longer run. This is particularly likely to happen in the case of product standards which create barriers to trade in the Internal Market.

Even if environmental laggard Member States are not principally opposed to the adoption of new legislation, they tend to prefer low environmental standards leaving little room for the Commission to positively influence environmental policy-making. While this conflicts with the Commission's interest in expanding its competencies, it is even more important from the perspective of the second main institutional interest which is characteristic for the Commission. This interest derives from the strong functional sectoralisation of the Commission as an organisation and, in fact, of the entire EU policy-making process (Peters, 1992, pp.117-119). As a result, there is intensive competition between different Directorate Generals, each of which wants to maximise its influence on EU policies.

EU environmental legislation is usually drafted by the Directorate General for Environment (DG Environment) which is generally sympathetic to environmental concerns and competes with other directorates for influence within the Commission (Mazey and Richardson, 1993, pp.121-122). Against this background Member State proposals for environmental legislation are more likely to be accepted by the Environment Directorate if they reflect a relatively high standard of environmental protection, and thus concord with its own preferences, rather than with those of competing directorate generals which frequently prefer lower standards.

The institutional structure of the Council of Ministers also offers opportunities for environmental leader Member States to push their legislative agenda. Leaders have used their respective Council presidencies to give their preferred policy proposals priority treatment on the Council's agenda (Sbragia, 1996, p.247). Of course, laggard countries may equally use the presidency to delay discussion of environmental legislation. However, on balance, the gains in decision-making efficiency accruing from the agenda-setting power of the presidency should be expected to reduce non-decision-making, and work in favour of the adoption of environmental legislation.

The functional differentiation of the Council provides leader Member States with particularly good opportunities to promote their policies. The Environment Council, which adopts most EU environmental legislation, is more sympathetic to environmental concerns than other Council formations or the General Council. Perhaps more importantly, environment ministries, who generally tend to be relatively weak vis-à-vis other national ministries, can use the Environment Council to outmanoeuvre rival ministries in a 'two-level game'. Given that rival national ministries are not represented on the Environment Council, environment ministries may promote strict environmental legislation at the EU level, which they were unable to push through at the national level due to resistance by other ministries. This explains why certain pieces of EU environmental legislation are stricter than any pre-existing national legislation (Sbragia, 1996, p.247).

The 'politics of expertise' is a final factor which often works in favour of environmental leader Member States. In drafting legislation the Commission is assisted by advisory committees which are usually staffed with experts from the Member States. Interactions in these committees are shaped by legal and technical discussions, rather than political ones (cf. Majone, 1994, pp.56-57). In the committee meetings representatives from leader Member States often dominate deliberations given that the environmental leaders tend to be more technically and economically advanced than laggard countries and have frequently already gained extensive experience with environmental regulation in the field under discussion (cf. Eichner, 1995). Consequently, representatives from leader states often have 'more to offer' to the Commission in terms of both task expansion and expertise (Andersen, 1997, p.222).

A similar logic applies to the institutional sub-structure of the Council of Ministers. Discussions in the Council Working Group on Environment, which prepares the meetings of the Council, often include technical experts from national administrations. In addition, *ad hoc* expert working groups may be established. Leader Member States, such as Denmark and Sweden, try to influence discussions in these bodies by providing well prepared substantive input (cf. Liefferink and Andersen, 1998, pp.261-262). Nevertheless, in the case of the Council the overall influence on policy outcomes of discussions at the more technical level has been found to be quite limited due to the dominant role of the more political bodies, in particular the Council itself, but also the Committee of Permanent Representatives (COREPER) (cf. Andersen, 1997, p.221).

On the whole, the development of European environmental policy has been surprisingly little affected by constellations of deadlock (see also Jordan, Brouwer and Noble, 1999b). Regulatory competition between environmentally progressive leader Member States and more conservative laggard countries has been a crucial factor in overcoming deadlock. The fact that Member States are not only motivated by their institutional self-interest in retaining competencies at the national level but also by a host of other ecological, political, economic and administrative factors increases the potential for reaching an agreement on common policies which exceeds the lowest common denominator and makes it possible to classify Member States as environmental leader, laggard or neutral countries.

In addition, leader Member States have disproportionate influence over EU environmental policy-making as a result of various institutional factors. The most important of these factors are the accessibility of the Commission to input from Member State officials, the institutionally set preferences of the Commission and its services, strong sectoralisation of EU decision-making and, to a lesser degree, opportunities to influence policy-making at the European level on the basis of expertise and other resources.

The evolution of an EU environmental policy, which clearly exceeds the lowest common denominator of Member State interests, may, however, have a price in terms of implementation problems. This consequence and other factors which account for the implementation deficit of EU environmental legislation, as well as some of the potential remedies that have been put in place, are discussed in the following section.

Problems of Implementation

There is broad agreement among analysts that EU environmental policy suffers from a serious implementation deficit. However, little is known about the exact extent of the deficit. This may be attributed to two general factors: First, there is no agreement on how to define and measure the implementation deficit. For example, while the Commission has developed detailed guidelines and indicators for the implementation of specific legal acts, including EU environmental legislation, it has not yet generalised the criteria which could determine when full implementation has been achieved. Consequently, it is difficult to compare and aggregate instances of implementation failure. One of the reasons for this is the fact that the Commission cannot interfere with the competencies of Member States, which are responsible for implementation (Nicolaides, 1999, pp.23-24). Second, there is a lack of information on the implementation of European legislation, in particular regarding practical implementation on the ground (McCormick, 1994, p.200; Collins and Earnshaw, 1992, p.236). When assessing implementation, the Commission focuses mainly on legal issues of transposition of European legislation into national legislation and on complaints from businesses and citizens about breaches of EU law (cf. EC, 1999, p.9; Nicolaides, 1999, p.24), rather than on more systematic methods of inquiry. This and other factors, such as the Commission's difficulties in assessing whether complaints about practical implementation are justified,[18] means that the data collected by the Commission provides only a very sketchy picture of the state of implementation of EU legislation (cf. Jordan, 1999a, pp.80-81).

Taking these limitations into account, the Commission's data nevertheless convey the impression that the implementation deficit is particularly large in the environmental field (or, at least, that the Commission perceives the implementation problems of EU environmental policy to be particularly significant). For example, in the Commission's 1999 Report on Monitoring the Application of Community Law (EC, 1999) the chapter on the environment takes up 37 pages. The second largest chapter – on Internal Market legislation – counts 35 pages, while all of the remaining chapters have less than 10 pages. These numbers are particularly striking if one takes into account that the number of applicable environmental directives was 145, whereas the corresponding number for the Internal Market – 745 – was almost five times higher. Although the number of applicable directives in most sectoral policies was significantly smaller than in the environmental field, it should also be pointed out that the Commission's report only contains five pages on legislation concerning agriculture despite the fact that there were 398 applicable directives in this field.

The number of times the European Commission decided to apply to the European Court of Justice (ECJ) for penalties for failure to implement EU environmental legislation also indicates that the Commission is particularly concerned about the implementation deficit of EU environmental legislation: of fourteen ECJ judgements regarding penalties given up to December 1998, nine concerned environmental directives (EC, 1999).

As mentioned above, implementation problems may occur at different stages of the process of implementing European law. First, EU directives might not be transposed correctly into national law (failure of legal transposition). Second, implementation might fail if the necessary administrative rules and institutional practices are not established (failure of administrative implementation). Finally, there might be a lack of technical, financial or personnel resources which hinders the full practical application and enforcement of EU legislation (failure of practical application).[19]

The various stages at which implementation problems occur suggest that the implementation deficit has several causes. For example, EU legislation may be difficult to implement because it is incoherent and/or vague, reflecting the political compromises reached by Member States during negotiations in the Council of Ministers. Incoherence and/or vagueness may lead to problems of legal transposition or difficulties in administrative implementation. EU legislation may also be difficult to implement if it ignores issues which are important at the stage of practical implementation. For example, there may be a lack of sufficiently qualified staff in some Member States or the costs of implementation for public or private actors may have been underestimated. This kind of 'mistake' can sometimes result from the fact that the Commission, which drafts EU legislation, is not responsible for implementation, which falls to the Member States alone (cf. Jordan, 1999a, pp.78-79). Majone (1994, p.66) argues that the Commission's regulatory activism may be explained in terms of this division of labour: Because the costs of implementation are borne by those who have to comply, rather than by the Commission, the latter has few incentives to regulate efficiently.

Implementation problems are frequently accompanied by a process of bargaining between the Commission and individual Member States because the 'Commission cannot command national or subnational actors, public or private' (Jordan, 1999a, p.78; see also Collins and Earnshaw, 1992, pp.228-229). Ultimately, referral of a case of implementation failure to the ECJ is politically costly and time-consuming. It is therefore only used as the Commission's last resort and many disputes are solved at an earlier stage through bargaining between the Commission and a Member State about the timing and extent of implementation.

Bargaining about implementation is likely to occur if Member States are unwilling or unable to fully implement EU environmental legislation because they do not support the aims of a particular piece of legislation, or if Member States are unwilling or unable to fulfil its administrative requirements, or if Member States lack resources. Against this background it has been argued that the fact that EU environmental legislation tends to reflect the preferences of environmental leader Member States may explain the implementation deficit. According to this view, implementation problems result from the fact that, as a consequence of the leader-laggard dynamic, the interests and administrative traditions of certain Member States are not sufficiently incorporated into EU environmental legislation. These countries are then more likely to cause implementation problems (cf. Börzel, 1999; see also Jordan, 1999, pp.76-77). For example, the southern Member States generally have a worse record of implementing EU environmental legislation than their northern partners. These countries frequently lack the economic and administrative resources which are needed for the practical application of the substantive and procedural environmental standards championed by leader Member States. In addition, domestic supporters of environmental protection, such as environmental non-governmental organisations (NGOs) or certain industries which gain from the application of strict standards, tend to be weak in these countries (cf. Börzel, 1999).[20]

In the late 1980s the implementation deficit of EU environmental legislation received growing attention from the Commission (Jordan, 1999a, pp.76-77) which embarked on a change of regulatory strategy. In this respect, the adoption of the 1992 Fifth Environmental Action Programme marked a turning point (cf. Heritier *et al.*, 1996, p.162). A legislative framework for European environmental policy was firmly in place and the Commission proclaimed a new phase of consolidating the legislative achievements. More specifically, to improve the implementation of EU environmental legislation, the Commission proposed involving national, sub-national and societal actors more strongly in the process of implementation (the principle of 'partnership'), improving the quality of, and access to, environmental information, simplifying legislation, and using more flexible regulatory instruments (cf. EC, 1996, p.3). These measures were expected to reduce the costs of implementing legislation while increasing the adaptability of EU legislation to local conditions and needs (Knill and Lenschow, 1999, p.597).[21]

In the following years several institutions were created which contributed to the implementation of the Fifth Environmental Action Programme. These institutions include the European Environment Agency (EEA) to improve the quality of information, the EPRG, mentioned above, the European Consultative Forum for the Environment and Sustainable Development, comprising various societal actors, and the Implementation Network for European Environmental Legislation (IMPEL) as a forum for national enforcement agencies.

The Commission also developed more flexible regulatory instruments based on procedural requirements and self-regulation. For example, the directives on access to environmental information and environmental impact assessment establish procedural requirements rather than substantive standards. Similarly, European legislation on environmental management and audit systems (EMAS) and eco-labels primarily aims at creating economic incentives for self-regulation. In addition, the Commission increasingly uses framework directives which are based on an integrated view of environmental problems. These directives allow for more flexible implementation, taking local conditions and interdependencies between different environmental problems into account. They also combine emission standards and the concept of Best Available Technique (BAT) – a more flexible version of the older concept of Best Available Technology – with quality standards and procedural requirements. The Integrated Pollution Prevention and Control (IPPC) Directive and the Water Framework Directive exemplify this approach (Scott, 2000; Hey, 2000; Matthews, 1999).

While it may still be too early to judge whether the Commission's new approach to regulation has improved the overall effectiveness of European environmental policy, original hopes that the new regulations would cause fewer implementation problems because of increased flexibility have so far been disappointed. Some Member States, such as Germany, have severe difficulties in implementing the new directives which conform more to traditional British than to German administrative practices. In addition, partly motivated by problems of administrative implementation, some Member States have abused the flexibility of the new regulations to circumvent or water down European requirements (cf. Börzel, 2000, p.35; Knill and Lenschow, 1999).

Although the implementation deficit was one of the primary reasons why the Commission outlined a new approach to EU environmental policy in the Fifth Environmental Action Programme,[22] other factors were relevant, too. The programme also reflected more fundamental concerns about the adequacy of traditional regulatory instruments to achieve sustainable development, about the costs of environmental regulations, and about the centralisation of decision-making in Brussels. In conjunction with the implementation deficit, these concerns gave rise to the programme's call for integration of environmental concerns into sectoral policies.

Integration of Environmental Concerns into Sectoral Policies

The compatibility of environmental protection with economic development and democracy was widely discussed at national and international levels in the late 1980s. The issue came up particularly forcefully in the context of the rise of global environmental problems, such as the depletion of the ozone layer, climate change, and the loss of biodiversity. It became increasingly necessary to consider how these problems could be solved without sacrificing economic development and democracy, in particular in developing countries. A preliminary answer to this question was given in the Brundtland Report, which introduced the concept of sustainable development (von Weizsäcker, 1990).

The Fifth Environmental Action Programme was entitled *Towards Sustainable Development*. It reflected these debates and applied them to the European context (Baker, 1997, pp.96-98). The existing approach of EU environmental policy was called into question by the realisation that a more preventive strategy was needed, allowing curative 'end-of-pipe' measures to be replaced by the integration of environmental concerns into the further development of the various sectors which cause environmental problems, such as industry, agriculture, transport etc. (cf. von Weizsäcker, 1990, pp.223-235). The extended political struggles of the 1980s over legislation to combat air pollution by prescribing end-of-pipe solutions – such as catalytic converters or desulphurisation technology (Boehmer-Christiansen and Skea, 1991; Holzinger, 1994) – had demonstrated that curative measures were simply too expensive to be used as the basis for a practicable and effective European approach to environmental policy. In particular for the economically and technologically less developed Member States the existing approach based on end-of-pipe technologies had proved politically untenable. The EU had therefore been forced to allow these countries to temporarily derogate from certain particularly expensive directives.[23] In addition to these measures, the Cohesion Fund was partly established as a re-distributive instrument to support the large investments in the environmental infrastructure of the four poorest Member States which were necessary to practically implement EU environmental legislation in these countries.

In addition to the relationship between environmental protection and economic development, the Fifth Environmental Action Programme also dealt with the issues of participation and democracy which had been raised in the context of the debate about sustainable development. In the EU discussion of these issues not only coincided with, but was also to some extent substantively linked to, political efforts to promote and apply the principle of subsidiarity. In the first half of the 1990s the UK, Germany and Denmark invoked the subsidiarity principle enshrined in the Maastricht Treaty to complain that the EU undermined the legitimate competencies of Member States and regions. On this view, environmental policy was dictated by Brussels and certain competencies should therefore be 'repatriated' (Collier, 1996, p.12).

Whereas some Member States used the subsidiarity principle to argue for a return of EU competencies to the national level, the Fifth Environmental Action Programme gives a different, more participatory interpretation of subsidiarity. It quotes Article A of the Maastricht Treaty, stating the aim of creating an ever closer union among the peoples of Europe, 'in which decisions are taken as closely as possible to the citizen' (EC, 1993, p.78). The Programme apparently took this aim to mean that the subsidiarity principle should be embedded within a broader concept of 'shared responsibility', according to which various state and societal actors, such as Member State authorities and regional governments, the business sector, the general public and consumers, should co-operate and participate in EU decision-making (cf. ibid. pp.78-79).

Despite the fact that sustainable development has become the official guiding principle of EU environmental policy and – according to Article 2 TEU and Article 2 TEC – is now one of the principal aims of the Union, only Article 6 TEC contains statements which may be read as a clearer (operational) definition of sustainable development (cf. Kraemer and Mazurek, 1999, pp.5-6; see also EC, 1998, p.3). Article 6 calls for the integration of environmental concerns into the formulation and implementation of sectoral policies to achieve sustainable development.

Against the background of what has been said above about sustainable development and the Fifth Environmental Action Programme, the integration principle of Article 6 may be interpreted as a requirement to promote integration at three different levels: in the societal and administrative spheres, and in the field of political decision-making.

Societal Integration

Sustainable development requires direct integration of environmental concerns into economic activities through technical and social innovation.[24] The EU's Auto-Oil Programme, which was launched in 1993 and is now in its second phase, exemplifies some aspects of societal integration. It takes a long-term approach to the problem of car emissions. The aim is to achieve a reduction of emissions by 70 per cent by 2010. The Auto-Oil Programme focuses on the establishment of a scientific basis for determining the extent to which emissions of pollutants should be reduced and on elaborating the most efficient ways to achieve the necessary reductions (cf. EC, n.d.)

Initially, the Environment, Industry and Energy Directorates General of the Commission and the European associations of car manufacturers (ACEA) and of the oil industry (EUROPIA) co-operated in the implementation of the programme. As there is a close interdependence between technologies to reduce car emissions, such as catalytic converters, and the improvement of fuel quality, participation by the respective industrial sectors was crucial. Additional actors – e.g. other relevant industries, NGOs, Member State representatives, and research institutes – have been included in the second phase of the programme (cf. Goodwin, 1998). Elaborating reduction strategies on the basis of scientific expertise and in co-operation with producers – other stakeholders function mainly as 'watchdogs' – the Auto-Oil Programme aims at ensuring that environmental concerns are fed into the long term technical development of the affected industries at an early stage.

Political initiatives to create a European Integrated Product Policy (IPP) also attempt to improve societal integration. The Commission has recently published a Green Paper on IPP (EC, 2001). According to the

Commission, IPP aims, among other things, at encouraging 'life-cycle thinking', stimulating consumer demand for environmentally friendly products, and establishing 'product panels' in which stakeholders work on ways to improve the environmental performance of products.

Administrative Integration

The second aspect of environmental policy integration is closely linked to implementation problems and flexible regulations.[25] The fact that the Commission increasingly uses framework directives which cover a wide range of environmental problems and regulatory instruments and allow for more flexible implementation reflects efforts to improve administrative integration. The IPPC Directive is a particularly important example because it seeks to integrate a wide range of diverse directives listed in its annexes into a single licensing procedure. It also combines different instruments, such as emission standards, BAT, quality standards, delegation of decision making, public information and participation. This broadly based approach aims at a better integration of sectoral economic, technical, local, and grass-roots concerns into administrative decision-making (cf. Scott, 2000).

If administrative integration is to enhance implementation it may often be necessary to simultaneously improve societal integration to create some of the pre-conditions for an effective implementation of flexible regulations. For example, the Auto-Oil Programme led to the adoption of new, significantly stricter but also more flexible EU regulations for cars and fuels. The 1998 Directive on the Quality of Petrol and Diesel Fuels, which resulted from the programme, contains temporally and substantively flexible requirements, allowing a Member State to delay full compliance if it can demonstrate that otherwise 'severe difficulties would ensue for its industries' (Art. 3). The directive also permits adoption of stricter standards in certain regions if 'atmospheric pollution [...] can reasonably be expected to constitute a serious and recurrent problem' (Art. 6 (1)).

Although more flexible environmental regulations may be more cost-efficient and better adapted to local environmental conditions, they have frequently been as difficult to implement as traditional regulations, mainly because of their considerable administrative requirements or the lack of domestic support for effective implementation by societal actors. Under these circumstances some Member States have, as mentioned above, abused flexible regulations to avoid implementing EU standards.

Against the background of these problems, the systematic integration of environmental concerns into the evolution of economic activities, outlined above, may complement flexible regulations in such a way that incentives are created which induce societal actors to support effective implementation. For example, the provisions of the Directive on the Quality of Petrol and Diesel Fuels allowing temporary derogations are more likely not to be abused if, as is the case, the Auto-Oil Programme signals to the oil industry in countries which invoke the temporary derogation that companies will be expected to comply with even stricter standards in the near future. Industry then has an incentive to use the time gained as a result of the temporary derogation to introduce the necessary changes.

Institutional Integration

The third dimension of integration of environmental concerns into sectoral policies concerns political decision-making structures. Reform of decision-making is a measure which will become necessary if less radical efforts to deal with environmental problems are insufficient. As mentioned in the introduction, despite the adoption of relatively strict EU environmental legislation, the overall state of the environment has not improved, and in many areas further deterioration is expected. For example, although the negative environmental impact of road traffic might decrease in the medium term as a result of, among other things, the Auto-Oil Programme, it will be necessary to embark on more radical, structural changes, including switching to more sustainable modes of transport, to prevent a reversal of this trend in the long term. The situation regarding CO_2 emissions caused by road traffic – an issue that is not being dealt with in the framework of the Auto-Oil Programme – is even more dramatic. In the short and medium term no improvements can be expected in this area, even if a whole range of planned measures were to be fully implemented (Goodwin, 1999).

Given the limited success of 'softer' instruments, institutional environmental policy integration has gained in importance in recent years. It has already been noted that the Amsterdam Treaty introduced Article 6 TEC, requiring the integration of environmental concerns into EU sectoral policies. Calling on the Transport, Agriculture and Energy Councils to prepare reports on the integration of environmental concerns into their respective fields of activity, the 1998 European Summit in Cardiff initiated the so-called 'Cardiff Process'. The three reports were presented in late 1999, together with additional reports prepared by several other Council formations. They received a very mixed evaluation from the Commission and will need further improvement (cf. EC, 1999).

Although the choice of instruments for achieving policy integration partly depends on the respective sector, the following instruments appear to be particularly significant: Ecological tax reform, the cutting of environmentally harmful subsidies, the introduction of liability rules for environmental damage, the 'strategic' environmental impact assessment of policies, programmes, and plans and, in particular, a general reform of political decision-making structures to overcome institutional barriers to integration and to create new possibilities for co-operation among decision-makers from different sectors.[26]

According to DG Environment fundamental changes will be necessary to achieve institutional integration, for example, in the transport sector and, above all, in the environment sector itself: 'The concept of integration will influence the role of different parts of the administration. It is not only the transport sector that has to adapt itself to the responsibility but also the environment sector will have a new role in assisting, pushing and monitoring the progress [of policy integration]' (EC, 1999a, p.8).

In the cases of societal and administrative integration some progress has already been made. In contrast, despite the growing political attention which the process of institutional integration has received in recent years, corresponding measures are mostly still waiting to be put into practice. The Commission's Communication on *Partnership for Integration* only contains several general proposals for the implementation of the integration principle of Article 6 TEC. In this document the Commission also mentions Agenda 2000 – which mostly deals with the implications of enlargement – as a major challenge for EU environmental policy that increases the need for institutional reforms to integrate environmental concerns into decision-making structures (EC, 1998, pp.8-9).

Taking the three fundamental characteristics of EU environmental policy-making as a yardstick, I now turn to the potential impact of enlargement on EU environmental policy and draw some preliminary conclusions with respect to integration of environmental concerns into sectoral policies, in particular transport policy.

The Impact of Enlargement

The enlargement of the Union to include, for the time being, ten CEECs (plus Cyprus and Malta) is often portrayed as a challenge to EU environmental policy. However, enlargement also offers important environmental opportunities (Carius *et al.*, 2000, pp.146-147). It is possible to argue that the challenge of enlargement for EU environmental policy lies to a large extent in the fact that it will be necessary to exploit the environmental opportunities which enlargement offers if additional problems for EU environmental policy are to be avoided.

The state of the environment in Central and Eastern Europe (CEE) is still characterised by a sharp contrast between heavily polluted so-called environmental 'hot-spots' and large, unspoilt areas possessing a rich biodiversity (EEA, 1998, p.149; REC, 1994, p.10). These particular features of the state of the Environment in CEE can mostly be attributed to the way in which the system of central planning operated in the era of communist rule. On the one hand, the population and industrial production, in particular heavy industries, were concentrated in a small number of regions close to cheap – and often dirty – sources of energy. On the other hand, many areas between these agglomerations remained relatively untouched. Similarly, although CEECs were plagued by severe industrial pollution, environmental degradation caused by environmentally harmful consumption patterns was relatively low.

However, since the political upheaval of the late 1980s and early 1990s there has been a clear trend towards convergence between the state of the environment in the EU and in CEECs. While this has brought considerable improvements to the environmental hot spots, convergence also means that 'CEECs could end up making the same mistakes as the west *after* the west has made them' (T&E, 2000). As economic growth in CEECs is expected to be significantly higher than in present EU Member States in the years ahead, it seems reasonable to assume that CEECs will increasingly contribute to those environmental trends which already pose the most challenging problems for EU environmental policy. Environmental degradation caused by road traffic is one of the relevant issues.

The challenge and the opportunity for EU environmental policy is, therefore, to limit, as far as possible, the tendency towards convergence in those areas where convergence results in a repetition of developments which are presently recognised in the EU as unsustainable. As pointed out above, the general framework of EU environmental policy-making in which many of the relevant decisions will be made is characterised by the leader-laggard dynamic, the implementation deficit, and the integration of environmental concerns into sectoral policies. Against this background, I argue, first, that to retain the dynamic development of EU environmental policy as a pre-condition for successful environmental policy-making and to prevent a growing implementation deficit after enlargement, it will be necessary to further increase the flexibility of EU environmental legislation. Second, such a strategy will only be successful in the longer run if the EU also increases its efforts to assist CEECs, in particular with respect to creating effective structures for the integration of environmental concerns into sectoral policies.

Regulatory Competition

If one looks at the factors which influence Member States' interest in EU environmental policy, one can expect that most CEECs will belong to the group of laggard Member States after accession. This holds, in particular, for economic and political factors. Although it is likely that economic growth rates in CEECs will be higher than in the present EU Member States, it will nevertheless take decades for the new Member States to catch up with the present members. This is particularly true for the least economically developed applicants, such as Bulgaria and Romania (Chalmers, 2000, p.24). The new Member States may therefore be expected to be less willing to bear the short and medium term costs of a high level of environmental protection.

In addition, despite the wave of environmental reforms in the early 1990s in most Accession Countries, today environmental protection ranks low on the political agenda in CEE. The economic and social problems of transition tend to dominate politics (Baker and Jehlicka, 1998, pp.9-11). Even the fact that several of the wealthier Accession Countries spend a significantly larger share of their GDP on environmental pollution abatement than most western European countries do (cf. OECD, 1999, p.130) offers little consolation for environmental leader Member States. This higher spending can be attributed, on the one hand, to particularly severe environmental problems caused by the legacy of environmental damage and a still relatively large share of unsustainable production technologies which these countries have inherited from the past and, on the other hand, to external pressure, in particular from the EU in the context of accession. The latter factor is underlined by the fact that the EU has repeatedly stressed that it expects Accession Countries not only to have completed the formal transposition of European environmental legislation into national law by the date of accession but, also, to be able to effectively apply and enforce the respective laws. The EU declared that it will only accept transitional periods for the practical implementation of EU environmental law in well-defined, exceptional cases.[27]

These circumstances suggest that with respect to economic and political factors, the situation in the Accession Countries resembles the one in the environmental laggard Member States Greece, Spain and Portugal at the time when these countries joined the EU. In some of the least developed Accession Countries the degree of economic development is in fact significantly lower than it was in the three southern Member States (Chalmers, 2000, p.24).

In the short and medium term, administrative problems also abound. The Accession Countries lack administrative capacities to implement and enforce European environmental legislation, in particular at the local and regional levels. However, in the longer run administrative capacity in the CEECs can be expected to improve. EU assistance, in particular through the PHARE institutional capacity building programme and the Accession Countries' own efforts to meet the requirements of the EU will lead to significant improvements. Perhaps more importantly, the CEECs are still going through a process of economic and political transformation. This ongoing process, coupled with the need to adapt to EU requirements, could create exceptional opportunities to overcome the institutional rigidities which usually tend to stifle administrative reforms (cf. Soil and Water Ltd., 1999, pp.61-69). Despite the lack of administrative capacities in Accession Countries, resistance to the introduction of more flexible regulations may therefore be weaker than in some of the old Member States, such as Germany. The fact that in the aftermath of the wave of political reforms in the early 1990s several CEECs adopted procedural regulations on public access to environmental information and environmental impact assessment – in some cases even 'strategic' environmental assessment of plans and policies – could also be helpful (cf. Caddy, 1998). In addition, most Accession Countries already made use of environmental charges and taxes in the 1970s and 1980s. It remains to be seen to what extent these traditions will in fact outweigh the lack of administrative capacities.

While administrative factors will, if anything, only contribute in the medium to long-term to the emergence of an interest of Accession

Countries in strict EU environmental legislation, ecological influences may be expected to do so sooner, albeit only in some Accession Countries. Regarding ecological factors, it seems useful to distinguish between those Accession Countries which are strongly affected by problems of transboundary pollution and border on present environmental leader Member States and the remaining Accession Countries for which these criteria do not apply. The first group is more likely to develop an interest in stricter EU environmental legislation than the second because it suffers more from transboundary pollution. However, even if transboundary pollution is not a serious problem, the fact that a particular country borders on an environmental leader Member State may create political incentives – for example, in terms of bilateral financial and technical assistance or more general efforts to improve the political climate – for this country to support a higher level of environmental protection than it would otherwise have opted for.

With the exception of Romania and Bulgaria, all Central and Eastern European Accession Countries have something approaching a common border with environmental leader Member States. Although Estonia, Latvia and Lithuania do not directly share a border with these countries, they do border on the Baltic Sea. Against the background that some environmental leader Member States, in particular Sweden and Denmark, pursue a very active environmental policy to protect the Baltic Sea (Haas, 1993), these countries may have a similar influence on the Baltic States as if they actually shared a border with them.

In addition, the ecological challenges which the Central and Eastern European Accession Countries have to confront are, in general, also more similar to those of the northern European leader Member States than to those of the southern laggards.[28] Unlike the southern Member States, Accession Countries have a long history of severe industrial environmental damage which may increase support from these countries for relatively strict environmental standards at the European level.

On balance, this short review of the factors which are likely to influence the interests of Accession Countries suggests that most Accession Countries would tend to oppose a high level of environmental protection once they became members of the EU. Economic, political and, at least in the short and medium term, administrative factors account for this hypothesis. Ecological aspects, in particular affectedness by transboundary pollution, geographical proximity to environmental leader Member States, and severe environmental problems which are qualitatively similar to those of the leader countries, may somewhat compensate for the impact of the negative political, economic and administrative factors.

If a group of countries which frequently adopts positions that are close to the positions of present environmental laggard Member States will join the EU, the leader-laggard dynamic may suffer. It would be significantly easier for a growing group of environmental laggard Member States to block EU legislation aiming at a high level of environmental protection. Laggard countries could also succeed more easily in getting lax environmental standards adopted at the European level. Finally, it would generally be more difficult to agree on decisions at all due to a further increase in the number and diversity of Member States as a result of enlargement (cf. Homeyer *et al.*, 2000).

Whether enlargement will seriously undermine the leader-laggard dynamic does not, however, entirely depend on the interests and the number of Member States represented in the Council. First, the impact of these factors on decision-rules may to some extent be mediated by a range of other possible influences, such as the increasing workload of the Council (cf. Golub, 2000). Second, environmental leader Member States frequently gain a disproportionate influence on EU environmental policy because they benefit from the institutional characteristics of the Commission and the availability of secondary institutional channels to promote their interests. Finally, new decision-making rules in the Council, as discussed for example at the 2000 Intergovernmental Conference, may partly compensate for the effects of rising Council membership if future reforms lead to an increase in the relative voting power of environmental leader Member States (Carius *et al.*, 2000, p.173)

Nevertheless, it seems difficult to prevent enlargement from complicating the functioning of the Council and undermining the leader-laggard dynamic. This is illustrated, among other things, by the fact that the number of EU members will increase by two thirds when all ten Central and Eastern European Accession Countries have joined the Union (Chalmers, 2000, p.24). In anticipation of these problems, the option of flexible integration has increasingly been discussed in recent years. More specifically, the Amsterdam Treaty introduced the possibility for a group of Member States to use the European institutions to engage in 'Closer Co-operation'. For example, according to Art. 11 TEC in conjunction with various other Treaty provisions, a majority of Member States may under certain conditions agree on common measures which are more far-reaching than existing Community legislation but are only binding for those Member

States which participate in that particular instance of co-operation. DG Environment has already considered the possibility that the provisions on Closer Co-operation may be usefully employed by environmental leader Member States to forestall stagnation of EU environmental policy as a result of enlargement (cf. EC, 1999b, p.46; see also Bär *et al.*, 1999).

Closer Co-operation could indeed preserve the leader-laggard dynamic after enlargement. Although those Member States which do not take part in a Closer Co-operation employing stricter environmental standards may in some cases enjoy an economic competitive advantage vis-à-vis the participants, the reverse may frequently also be true. For example, the members of a Closer Co-operation may become technological and institutional pace-setters which have the opportunity to unilaterally determine the standards and procedures with which the non-members would have to comply if they decided to increase their level of environmental protection. In this case the 'outs' would have little choice but to join the Closer Co-operation whose standards and procedures would already be firmly entrenched at the EU level by the Commission and the founding members of the Closer Co-operation.

In other cases incentives for the non-members to adopt the higher standards developed in the framework of Closer Co-operation may be permanently too weak. However, even in these cases there appears to be a high probability that the 'outs' would eventually join the Closer Co-operation. First, the Commission has a strong interest in preventing permanently differing standards between groups of Member States and may therefore propose measures to assist the laggards in catching up. Second, Member States already participating in a Closer Co-operation may also have an interest in offering assistance and incentives for non-members to join, in particular if joining would contribute to the elimination of an economic competitive advantage for the non-members. Finally, in certain cases, for example if it seems likely that the members of a Closer Co-operation would benefit from pace-setter advantages, environmental leader Member States may use the option of establishing a Closer Co-operation merely as a threat to induce laggard countries to agree to higher standards.

Implementation Problems

Not only does enlargement threaten to undermine the leader-laggard dynamic, but it may also increase the implementation deficit of EU environmental legislation. The findings of the Commission's 1999

Screening of the compatibility of the environmental legislation of Accession Countries with EU requirements (see for example, EC, 1999c), the Commission's annual reports on progress towards accession (see for example, EC, 1999d), and the documents prepared for the accession negotiations in the field of the environment (see for example, Government of the Czech Republic, 1999; Council of the European Union, 1999) highlight many implementation problems. Severe difficulties exist at each stage of implementation. In most Accession Countries the process of legal transposition has been significantly slower than expected. More importantly, the Accession Countries lack administrative and financial resources to effectively apply and enforce environmental legislation on the ground. If the first five Accession Countries joined the EU according to their own ambitious plans in 2003, most of them would need several long transitional periods of five, ten, or even more years until they were capable of fully applying and enforcing central pieces of EU environmental legislation, in particular the Urban Waste Water Treatment Directive (UWWTD), several other water related directives, the IPPC Directive, and various items of waste legislation (Homeyer *et al.*, 1999).

There are at least three underlying problems which give rise to the need for an exceptionally large number of transitional arrangements in the environmental field. First, many Accession Countries lack sufficiently effective institutions to implement and enforce environmental legislation. This holds, in particular, for environmental inspectorates responsible for monitoring and enforcement, local and regional authorities, and in some cases even environment ministries themselves. These institutions suffer from a lack of resources, such as qualified staff and technical equipment, and from inefficient institutional arrangements (OECD, 1999, p.64; EEA, 1999, p.401). Frequently, financial requirements for the practical application of EU environmental legislation constitute a second reason why transitional periods are needed. In 1997 the respective costs were estimated at a total of 120 billion Euro over 20 years for all Central and Eastern European Accession Countries (EPE, 1997, pp.18;97). Less than ten environmental directives dealing with water and air pollution and with waste disposal have been identified as being responsible for most of the heavy investment needs. Among these, practical implementation of the UWWTD is by far the most costly requirement (Soil and Water Ltd., 2000, pp.40; 70-71). However, it should be kept in mind that the costing assessments prepared so far are fraught with methodological difficulties and differ in their findings (Carius *et al.*, 2000a; Soil and Water Ltd., 2000, p.63).

Implementation is also hindered by the fact that environmental NGOs and other societal actors which could be expected to politically support full implementation of EU legislation are relatively weakly developed in most Accession Countries (cf. OECD, 1999, pp.81-102). Although this is not directly relevant for fulfilling EU requirements, it is nevertheless important for assessing future implementation problems. For example, similar weaknesses contribute to implementation problems of EU environmental legislation in some of the southern laggard Member States. In these countries the practical application of many environmental directives suffers from the fact that there is little political support from domestic societal actors (cf. Börzel, 1999).

Integration of Environmental Concerns into Sectoral Policies

The interest of environmental leader Member States and the Commission in preserving the leader-laggard dynamic and, perhaps more importantly, the need to prevent any further increase in the implementation deficit as a consequence of enlargement is likely to lead to an increasing application of forms of more flexible integration along the lines of Closer Co-operation and flexible regulation (Homeyer *et al.*, 2000). This increases the risk that permanent differences in the level of environmental protection will emerge among Member States. It may also create better opportunities for Member States to 'cheat' in the process of implementing flexible EU environmental regulations. Promoting integration of environmental concerns into sectoral policies offers the chance of counteracting these tendencies. As a result of policy integration, societal and political actors in 'non-environmental' sectors may gradually develop an interest in, and the capacity to promote, environmental protection.

Unfortunately, given the limited organisational and administrative resources in the Accession Countries at societal and political levels, co-ordination between different actors to achieve policy integration may be particularly difficult to realise in these countries. In addition, Accession Countries frequently have a long tradition of bureaucratic policy-making which is characterised by intense rivalries among and between ministries and branches of government, rendering inter-sectoral co-operation even more difficult (Carius *et al.*, 2000a).

Nevertheless, the process of joining the EU, in conjunction with the more general ongoing transition process in CEECs, may have opened a window of opportunity to promote the integration of environmental concerns into sectoral policies. First, as pointed out above, the transition and accession processes tend to reduce the institutional rigidities which frequently hinder successful institutional reform. Second, the process of joining the EU has led to the formation of new intra- and inter-sectoral co-ordination structures. For example, Slovenia has established an Office for European Integration which co-ordinates and monitors the transposition and implementation of EU legislation by the various ministries (cf. ECE, 1999, p.3). Similar structures have been created in other Accession Countries. In addition to a central co-ordinating body, they usually include a European integration unit in the ministries which are most affected by the accession process. It may be possible for Accession Countries to build on their experiences in co-ordinating the accession process and to design and implement measures to improve environmental policy integration. After accession when they have lost their original function, it may even be possible to use the existing co-ordination structures for this purpose.

Third, the accession process provides the EU with the means to exert exceptionally strong influence on institutional reforms in the Accession Countries (cf. Grabbe, 1999) which may be used to promote the integration of environmental concerns into sectoral policies. This opportunity appears to be particularly remarkable given that the fact that the Member States, rather than the Commission, are responsible for implementation seems, at least in part, to account for the current implementation deficit. More specifically, the Commission has a much larger influence on how EU environmental legislation is implemented in Accession Countries than it normally has on implementation in the Member States, because it has been charged with assessing the progress made by Accession Countries in the approximation of EU legislation and in implementing the Union's pre-accession strategy which supports these countries in adopting and implementing EU legislation. The Commission's influence is further increased by a political power differential between the EU and Accession Countries which is caused by the fact that it is the Commission and, ultimately, the present Member States which eventually decide when, and under which conditions, a country may join the Union.[29] Under these circumstances 'the European Commission has taken the lead in defining the *acquis communautaire* for CEE' (Grabbe, 1999, p.24).

Despite these opportunities, the EU so far appears not to have given priority to achieving policy integration in the framework of the accession process. First, this is exemplified by the approach to negotiating transitional periods. According to the Commission, transitional arrangements may be justified on political or economic grounds. For example, transitional periods may be agreed where huge investments are needed, direct payments are to be made to farmers under the CAP, or where the free movement of persons within the Union is concerned. By contrast, transitional periods will not usually be granted for difficulties of administrative implementation. With a view to the environmental sector there are, however, at least two reasons why transitional arrangements for problems of administrative implementation should also be considered admissible.

First, although Accession Countries will probably be able to fully transpose EU legislation into national law before accession, they may continue to lack the administrative capacity to practically apply some of the administratively more challenging pieces of EU environmental legislation. If the EU refuses to consider transitional periods in these cases, Accession Countries would have little choice but to erect 'Potemkin-village' organisational structures to prevent further delays in the accession process. These structures would allow the EU to maintain that the requirements for accession had been fulfilled although, in fact, the Accession Countries would still not have the administrative capacities necessary for practical application. Accession Countries frequently try to get *ex ante* approval from the Commission for specific ways of implementing EU legislation. These efforts may often reflect the wish to minimize *ex post* monitoring by the Commission which would threaten to reveal the janus-face of certain administrative arrangements (Jacoby, 1999).

If transitional periods were considered for administratively challenging directives, the incentives and opportunities for this kind of 'cheating' could be reduced. On the one hand, the pressure on Accession Countries to demonstrate full compliance would be less severe because they could request more transitional periods. On the other hand, the Commission could retain its special competencies for monitoring and supporting implementation beyond the date of accession in those cases in which transitional periods were granted. The long term risk of an increasing implementation deficit following enlargement might be significantly reduced as a result.

Second, admitting transitional periods for administratively challenging legislation may also help to implement the integration principle of Art. 6

TEC and similar statements in the Fifth Environmental Action Programme. The Commission has relatively few means of putting pressure on Accession Countries to implement these quite general requirements before the date of accession and it is likely to have even less influence once the Accession Countries have become regular Member States. However, in those areas where transitional periods are granted, the Commission could retain more influence on the process of implementation beyond the date of accession. As with implementation in general, this would give the Commission additional opportunities to promote the integration principle.

The way in which the Commission has handled the EU's pre-accession strategy and its assistance programmes to support the Accession Countries also suggests that it has so far failed to press for better environmental policy integration in the Accession Countries. In theory, the PHARE programme could be used as an instrument to strengthen the integration of environmental concerns into sectoral policies. Thirty per cent (about 500m Euro) of the annual PHARE budget is devoted to institutional capacity building in the Accession Countries, but environmental policy integration has so far not been a priority. Similarly, environmental concerns have only been weakly integrated into the provisions of the Instrument for Structural Policies for Pre-Accession (ISPA) Regulation which govern EU support for major infrastructure investment in the transport sector in the Accession Countries. For example, the ISPA Regulation contains weaker provisions on environmental safeguards than the corresponding rules for the EU's Cohesion Fund on which ISPA was modelled. In addition, unlike the Cohesion Fund, ISPA only supports large projects.[30] Environmentally friendly, smaller and less capital intensive investments are therefore not supported by ISPA.

Finally, the Transport Infrastructure Needs Assessment (TINA), which is the Commission's main planning instrument for transport infrastructure investment in the Accession Countries, so far seems to have been subjected to less environmental impact assessment than the Trans-European Network for Transport (TEN-T), the corresponding measure for the present Member States, although the Commission has declared its intention to subject TINA to a full blown 'strategic' environmental impact assessment (Fergusson, 2000, pp.5-6).

These shortcomings are particularly deplorable because the transport sector in CEE still exhibits several characteristics which are favourable from the point of view of sustainable development and might be preserved if environmental policy integration at national and EU levels was improved. For example, the modal split between rail and road transport is much more balanced in Accession Countries than in the present Member States. If the integration of environmental concerns into European and national transport policies is not significantly improved, the situation in CEE will most likely continue to evolve rapidly towards western European conditions (ibid., pp.1-2).

Conclusion

This chapter argued that EU environmental policy exhibits three major characteristics: the leader-laggard dynamic, the implementation deficit, and initiatives to integrate environmental concerns into sectoral policies. The first of these factors accounts for much of the dynamic development of EU environmental legislation over almost thirty years. To some extent the second factor – the implementation deficit – actually reflects the success of EU environmental policy at the legislative level. Frequently, individual Member States are unwilling or unable to keep pace with the dynamic development of EU environmental legislation. Although the Commission adopted a more flexible regulatory approach in the 1990s, it has not yet been possible to significantly reduce the implementation deficit. In some cases flexible regulations have in fact caused their own implementation problems. Third, despite the proliferation of EU environmental legislation, the overall state of the European environment has not improved. On the contrary, growing problems associated with climate change, waste disposal, increasing land use, and loss of biodiversity suggest that the overall situation may be deteriorating. Against this background, the need to integrate environmental concerns into sectoral policies is receiving increasing political attention. Above all, environmental policy integration may help to bring about the structural changes which are needed to reverse the trend towards environmental degradation. In addition, it may lead to better implementation of flexible regulations, which is likely to increase the effectiveness of, and the political support for, EU environmental policy.

EU enlargement will create further pressure to put the environmental integration principle into practice. First, enlargement threatens to undermine the leader-laggard dynamic because most Accession Countries will probably oppose relatively strict environmental standards once they have joined the EU. Environmental leader Member States may therefore increasingly make use of mechanisms of flexible integration, for example the Treaty provisions on Closer Co-operation. This development could preserve the leader-laggard dynamic but might also create new problems. Second, enlargement also threatens to further increase the implementation deficit of EU environmental legislation. No matter whether the first round of enlargement takes place in 2003 or, as seems more likely, a few years later, Accession Countries lack the financial resources and administrative capacities to achieve full practical implementation of EU environmental legislation by the date of accession. This may lead to a further increase in regulatory flexibility on the part of the EU. Finally, integration of environmental concerns into sectoral policies promises to reduce the risks inherent in increasingly flexible integration and regulation. On the one hand, there is a risk that groups of Member States characterised by permanently different levels of environmental protection might emerge. On the other hand, flexibility is likely to create better opportunities for Member States to 'cheat' in the process of implementing EU environmental legislation. Although better environmental policy integration may help to reduce these risks, the Commission has, as yet, made few efforts to promote policy integration in the framework of the pre-accession strategy.

If present efforts at the EU level to integrate environmental concerns into sectoral policies were successful and could be transferred to the Accession Countries, it might be possible to preserve some of the environmentally favourable conditions in CEE, such as a relatively high level of biodiversity and an environmentally friendly split between alternative modes of transport. There are at least two reasons why the conditions for promoting environmental policy integration in the Accession Countries would appear to be favourable at the moment: First, institutional and economic structures in these countries are currently relatively malleable due to the ongoing processes of transformation and EU accession. Second, against the background of a strongly asymmetric distribution of power in favour of the EU during the process of enlargement, the EU can exert exceptionally strong influence on the Accession Countries in the framework of the special pre-accession regime. In view of these circumstances it might be possible to significantly strengthen environmental policy integration in Accession Countries were this aim to become a high priority of the EU accession strategy.

Notes

1   The term 'European Union' will be used in this chapter for reasons of simplicity, although in several cases the term 'European Community' (EC) would be the correct one from a historical or legal perspective.

2   In addition there are three non Central and Eastern European Accession Countries: Cyprus, Malta and Turkey. In contrast to the other candidates, accession negotiations have not yet been formally initiated with Turkey. Croatia is another CEEC which may become a candidate for accession in the next few years.

3   See also Andersen (1997).

4   The ambiguous overall record of EU environmental policy which results from implementation problems had already been noted in the early 1990s. See, for example, Liberatore (1991).

5   See also Knill and Lenschow (1999), p.592.

6   If anything, Scharpf's model of the joint decision trap applies to EU environmental policy in the 1970s. Cf. Jordan (1999), p.7; Andersen (1997), p.212.

7   For a comparison of different theoretical explanations for EU 'task expansion' in the environmental field, see Zito (1999) and Andersen (1997), pp.217-224.

8   Andersen (1997, pp.222-224) stresses the particular importance of competitive interaction between leaders and laggards. According to Zito (1999), EU 'task expansion' in the environmental field can mostly be explained in terms of Member State interests and the European Commission's influence on policy-making, although other explanations are relevant, too. He argues that more empirical research is needed to establish the overall explanatory weight of accounts in terms of the influence of Member States vis-à-vis explanations which emphasize the independent role of the Commission. It should be stressed that in many cases explanations may be complementary, rather than mutually exclusive. If this is the case, the question is not so much *how* the respective event can be explained, but *how much* can be explained by each of the various accounts and to what extent they may be combined. Cf. Knill and Lenschow (2000, p.6), Mayntz and Scharpf (1995), p.52; Peterson (1995).

9   For a short overview, see Holzinger (1994), pp.76-77.

10   For a similar classification of Member States' interests in European environmental policy, see Liberatore (1991), pp.286-289.

11   Product standards apply to the product, whereas process standards apply to the production process.

12   In addition, high product standards tend to prevail if consumers prefer, are willing to pay for, and are able to recognise, high-quality products. Cf. Scharpf (1998), pp.126-127.

13   Production standards do not directly affect the Internal Market. Therefore it is easier for laggard Member States to push for EU regulations which reflect the lowest common denominator or to block any efforts at harmonisation. Given a liberalised Internal Market, this subjects the process of adjustment of production standards to strong pressures of economic competition. Recent debates about the adoption of carbon/energy taxes in the EU – which can be broadly treated as process standards – illustrate this point, although the respective Commission proposal eventually failed for a different reason, e.g. the institutional interest of Member States in preventing

the EU from gaining additional competencies in the area of taxation. See Delbeke and Bergmann (1998), pp.244-245.

14   Of course, this argument ignores the fact that *other countries* may suffer from transboundary pollution originating in the U.K.

15   Heritier argues that the U.K. not only softened its position regarding EU environmental policy in the late 1980s but also acted as a leader Member State in bringing about certain changes in the administrative requirements of EU environmental legislation. See Heritier (1995), p.279.

16   In theory, the restrictions posed by unanimous decision-making could also be overcome by issue-linkage. However, in the day-to-day EU legislative process cross-sectoral issue-linkage is rarely practised due to the strong functional sectoralisation of decision-making. Intra-sectoral issue linkage is more common but also less effective due to more limited opportunities for exchange. See Golub (2000), pp.4-5, and Golub (1996), p.710.

17   At the highest level, even Commissioners play an important role in promoting the interests of 'their' Member States. For example, German Environment Commissioner Narjes promoted the German interest in the adoption of EU legislation to combat air pollution and Danish Commissioner Bjerregard represented the interests of smaller Member States. Cf. Andersen (1997), pp.213; 216.

18   For example, on complaints regarding the Directive on environmental impact assessment (85/337/EEC) the Commission comments that 'it is obviously difficult for Commission departments to investigate cases where the quality of impact assessment is questioned or it is contended that their findings are not properly acted upon. Although the Directive contains Articles regarding the content of impact assessments, it is difficult to verify the compliance with them by national authorities; moreover, it is not always easy to contest the merits of a choice taken by national authorities'. See European Commission (1999), p.11.

19   There are different conceptions of how many principal steps are involved in the implementation of EU legislation. For example, according to the Commission there are four steps: Legal transposition, practical application, compliance, and enforcement (European Commission, 1997a). Evaluation and policy reform may be added to this list (Nicolaides, 1999, p.5). Others distinguish merely between the two stages of legal transposition and practical implementation (Liberatore, 1991, pp.298-299). The threefold distinction, presented above, has the advantage that it explicitly deals with the question of resources which is particularly relevant for the effects of enlargement on EU environmental policy.

20   Despite the fact that the UK is often regarded as an environmental laggard Member State, it has a relatively favourable record of implementing EU environmental legislation. This may partly be explained by the fact that the U.K., which is one of the four large Member States, has better chances of influencing EU decision-making than smaller laggard countries. As noted above, in the late 1980s the U.K. has sometimes even acted as a leader country and has managed to export its regulatory approach to the EU. In addition, the U.K. has more resources to implement policies than less wealthy laggard countries. By contrast, the implementation record of environmental leader Member State Germany is mixed. This may partly be explained by economic difficulties following German unification and by the fact that some of the more recent environmental directives conflict with the traditional German regulatory approach. Cf. Knill and Lenschow (1999).

21   On the advantages of more flexible EU environmental regulation, see also Holzinger (1999).

22   As Liberatore (1991) shows, many of the concepts and ideas presented in the Fifth Environmental Action Programme had at least partly been contained in previous action programmes. However, it was only in the early 1990s that the political context was such that considerable efforts were made to put these concepts into practice (cf. Jordan, 1999b, p.16).

23   For example, the Large Combustion Plants Directive grants Spain a special transitional period for compliance with important requirements. In addition, Art. 15 of the Single European Act allows for transitional periods if certain economies are disproportionately burdened by measures to create the Internal Market. Cf. Ehlermann (1995), and Beck (1995), p.150.

24   Societal integration of environmental concerns seems to be particularly advanced in the Netherlands. See Jordan and Lenschow (2000), pp.114-115.

25   Among EU Member States, Denmark is relatively advanced in using fiscal instruments to achieve environmental policy integration in a cost efficient, flexible way. Cf. Jordan and Lenschow (2000), pp.115-116.

26   For a discussion of these and other instruments see, for example, Buck *et al.* (1999); European Environment Agency (1999), pp.399-406; Kraemer and Mazurek (1999).

27   Recent statements to this extent can be found in the European Union's Common Positions for the accession negotiations in the field of the environment. See, for example, Council of the European Union (1999), pp.1-2.

28   On the different environmental challenges for northern and southern European Member States, see Holzinger (1994), pp.76-77

29   Although EU Member States ultimately decide on accession, the long-drawn out pre-accession process of monitoring and influencing the adoption and implementation of EU legislation in the Accession Countries is a highly technical exercise which is dominated by the Commission. For the case of environmental legislation, see Homeyer *et al.* (2000) and Caddy (1997). More generally, see Jacoby (1999) and Grabbe (1999a).

References

Albin, S. and Bär, S. (1999), 'Nationale Alleingänge nach dem Vertrag von Amsterdam. Der neue Art. 95 EGV: Fortschritt oder Rückschritt für den Umweltschutz', *Natur und Recht*, 21(4), pp.185-192.

Andersen, M.S. (1997), 'Environmental Policy in Europe', in Hesse, J.J. and Toonen, T.A.J. (eds), *European Yearbook of Comparative Government and Public Administration*, Vol. 3, Nomos, Baden-Baden, pp.205-226.

Baker, S. (1997), 'The Evolution of European Union Environmental Policy. From Growth to Sustainable Development?', in Baker, S., Kousis, M., Richardson, D. and Young, S. (eds), *The Politics of Sustainable Development. Theory, Policy and Practice within the European Union*, London, New York, Routledge, pp. 91-106.

Baker, S. and. Jehlicka, P. (1998), 'Dilemmas of Transition: The Environment, Democracy and Economic Reform in East Central Europe – An Introduction', in Baker, S. and Jehlicka, P. (eds), *Dilemmas of Transition: The Environment, Democracy and Economic Reform in East Central Europe*, Frank Cass, Ilford, pp.1-26.

Bär, S., Homeyer, I.V., Kraemer, A., Mazurek, A.-G., and Klasing, A. (1999), *Closer Co-operation in European Environmental Policy after Amsterdam*, Schriftenreihe des BMUJF, Band 32/1999, Austrian Federal Ministry of Environment, Youth and Family, Vienna.

Beck, H. (1995), *Abgestufte Integration im Europäischen Gemeinschaftsrecht unter besonderer Berücksichtigung des Umweltrechts*, Europäische Hochschulschriften, Peter Lang, Frankfurt/Main.

Boehmer-Christiansen, S. and Skea, J. (1991), *Acid Politics: Environmental and Energy Policies in Britain and Germany*, Belhaven Press, London.

Börzel, T.A. (1999), 'Why there is no Southern Problem. Environmental Leaders and Laggards in the European Union', *EUI Working Paper* RSC No. 99/16, European University Institute, Florence.

Börzel, T.A. (2000), 'Best Practice Solution or Problem for the Effectiveness of European Environmental Policy?', *EUI Review* Spring 2000, European University Institute, Florence, pp. 32-36.

Buck, M., Kraemer, R.A. and Wilkinson, D. (1999), 'Der Cardiff-Prozess zur Integration von Umweltschutzbelangen in andere Sektorpolitiken', *Aus Politik und Zeitgeschichte* B 48/99, pp.12-20.

Caddy, J. (1997), 'Harmonization and Asymmetry: Environmental Policy Co-ordination Between the European Union and Central Europe', *Journal of European Public Policy* Vol. 4, No. 3, pp.318-360.

Caddy, J. (1998), *Sowing the Seeds of Deliberative Democracy? Institutions for the Environment in Central Europe: Case Studies of Public Participation in Environmental Decision-making in Contemporary Hungary*, Ph.D. Dissertation, European University Institute, Florence.

Carius, A., Homeyer, I.V. and Bär, S. (2000), 'The Eastern Enlargement of the European Union and Environmental Policy: Challenges, Expectations, Multiple Speeds and Flexibility', in Holzinger, K. and Knoepfel, P. (eds), *Environmental Policy in a European Union of Variable Geometry? The Challenge of the Next Enlargement*, Helbing and Lichtenhahn, Basel, pp.141-180.

Carius, A., Homeyer, I.V., Bär, S. and Kraemer, A.R. (2000a), *Die umweltpolitische Dimension der Osterweiterung der Europäischen Union: Herausforderungen und Chancen*, Metzler-Poeschel, Stuttgart.

Chalmers, M. (2000), *Paying for EU Enlargement – Can a New "Burdensharing Bargain" be Sustained?*, Paper presented at the Political Studies Association-UK 50th Annual Conference, 10-13 April 2000, London.

Collier, U. (1996), 'Deregulation, Subsidiarity and Sustainability: New Challenges for EU Environmental Policy', *EUI Working Paper* RSC No. 96/60, European University Institute, Florence.

Collins, K. and Earnshaw, D. (1992), 'The Implementation and Enforcement of European Community Environment Legislation', *Environmental Politics*, Vol. 1, No. 4, pp.213-249.

Council of the European Union (1999), *European Union Common Position. Chapter 22: Environment, Conference on Accession to the European Union – Czech Republic, CONF-CZ 63/99*, 30 November 1999, Council of the European Union, Brussels.

Delbeke, J. and Bergmann, H. (1998), 'Environmental Taxes and Charges in the EU', in Golub, J. (ed.), *New Instruments for Environmental Policy in the EU*, Routledge, London, pp.242-260.

Den Boer, M., Guggenbühl, A. and Vanhoonacker, S. (eds) (1998), *Coping with Flexibility and Legitimacy after Amsterdam*, European Institute of Public Administration, Maastricht.

Economic Commission for Europe (1999), *EPR of Slovenia: Report on Follow-Up, Environmental Performance Reviews*, United Nations, Geneva.

EEA, European Environment Agency (1998), *Europe's Environment: The Second Assessment*, Office for Official Publications of the European Communities, Luxembourg.

Ehlermann, C.D. (1995), 'Increased Differentiation or Stronger Uniformity', *EUI Working Paper RSC No.* 95/21, European University Institute, Florence.

Eichner, V. (1995), 'European Health and Safety Regulation: No "Race to the Bottom"', in Waarden, F. van and Unger, B. (eds), *Convergence or Diversity? The Pressure of Internationalization on Economic Governance Institutions and Policy Outcomes*, Avebury, Aldershot, pp.229-251.

ENDS Daily (2000), 'EU Accession Countries Win More Green Funds', *ENDS Daily*.

Environment Policy Europe (1997), *Compliance Costing for Approximation of EU Environmental Legislation in the CEEC*, Environment Policy Europe, Brussels.

European Commission (1993), *Towards Sustainability. Fifth Environmental Action Programme*, COM(94) 465, Office for Official Publications of the European Communities, Brussels.

European Commission (1996), *Implementing Community Environmental Law*, COM(96) 500 final, European Commission, Brussels.

European Commission (1997), *Agenda 2000*, COM(97) 2000, European Commission, Brussels.

European Commission (1997a), *Commission Staff Working Paper – Guide to the Approximation of European Union Environmental Legislation*, European Commission, Brussels.

European Commission (1998), *Partnership for Integration – A Strategy for Integrating Environment into European Union Policies*, COM(98) 333, European Commission, Brussels.

European Commission (1998a), *Accession Strategies for Environment: Meeting the Challenge of Enlargement with the Candidate Countries in Central and Eastern Europe*, COM(98) 294, European Commission, Brussels.

European Commission (1999), *Environment Chapter of the 16th Annual Report on Monitoring the Application of Community Law (1998)*, COM(99) 301, European Commission, Brussels.

European Commission (1999a), *From Cardiff to Helsinki and Beyond. Report to the European Council on Integrating Environmental Concern and Sustainable Development into Community Policies*, SEC (1999) 1941 final, Commission Working Document, European Commission, Brussels.

European Commission (1999b), *The Global Assessment: Environment and Sustainable Development Policy Beyond 2000*, Draft at 16 July 1999, DG Environment, European Commission, Brussels.

European Commission (1999c), *Czech Republic – Screening Results. Chapter 22 – Environment, Enlargement*, European Commission, Brussels.

European Commission (1999d), *1999 Regular Report from the Commission on Czech Republic's Progress towards Accession*, European Commission, Brussels.

European Commission (2001), *Green Paper on Integrated Product Policy*, COM(2001) 68, European Commission, Brussels.

European Commission (n.d.), *The European Auto Oil Programme*, A Report by the Directorate Generals for: Industry; Energy; and Environment, Civil Protection & Nuclear Safety, European Commission, Brussels.

European Commission (n.d.a), *Sector Integration Process – Towards the Helsinki Summit*, Background Paper, DG Environment, European Commission, Brussels.

European Environment Agency (1999), *Environment in the European Union at the Turn of the Century*, Office for Official Publications of the European Communities, Luxembourg.

Fergusson, M. (2000), 'Transport, Enlargement and the Environment', Background Paper to the Transport and Environment, Conference on Transport and EU Enlargement, European Federation for Transport and Environment, Brussels.

Golub, J. (1996), 'British Sovereignty and the Development of EC Environmental Policy', *Environmental Politics*, Vol. 5, No. 4, pp.700-728.

Golub, J. (1996a), 'State Power and Institutional Influence in European Integration: Lessons from the Packaging Waste Directive', *Journal of Common Market Studies*, Vol. 34, No. 3, pp.313-339.

Golub, J. (1998a), 'New Instruments for Environmental Policy in the EU. Introduction and Overview', in Golub, J. (ed.), *New Instruments for Environmental Policy in the EU*, London, Routledge, pp.1-29.

Golub, J. (2000), *Institutional Reform and Decision-Making in the European Union*, Paper presented at the Political Studies Association-UK 50th Annual Conference, 10-13 April 2000, London.

Golub, J. (ed.) (1998), *New Instruments for Environmental Policy in the EU*, Routledge, London.

Goodwin, F. (1998), *Auto Oil Briefing Note*, European Federation for Transport and Environment, Brussels.

Goodwin, F. (1999), 'Response to the Commission Communication on the Common Transport Policy: Perspectives for the Future', *Transport and Environment* 99/5, European Federation for Transport and Environment, Brussels.

Government of the Czech Republic (1999), *Negotiating Position of the Czech Republic on Chapter 22 Environment*, Intergovernmental Conference on the Accession of the Czech Republic to the European Union, 14 July 1999, Brussels.

Grabbe, H. (1999), 'The Transfer of Policy Models from the EU to Central and Eastern Europe: Europeanisation by Design?', Paper presented at the 1999 Annual Meeting of the American Political Science Association, 2-5 September 1999, Atlanta.

Grabbe, H. (1999a), 'A Partnership for Accession? The Implications of EU Conditionality for the Central and East European Applicants', *EUI Working Paper* RSC No. 12/99, European University Institute, Florence.

Haas, P.M. (1993), 'Protecting the Baltic and North Seas', in Haas, P.M., Keohane, R.O. and Levy, M.A. (eds), *Institutions for the Earth. Sources of Effective International Environmental Protection*, MIT Press, Cambridge, pp.133-181.

Heritier, A. (1995), 'Leaders and Laggards in European Clean Air Policy', in Waarden, F. van and Unger, B. (eds), *Convergence or Diversity? The Pressure of Internationalization on Economic Governance Institutions and Policy Outcomes*, Aldershot, Avebury, pp.278-305.

Heritier, A. (1998), *Second-Order Institutionalisation in Europe: How to Solve Collective Action Problems Under Conditions of Diversity*, Preprint from the Max-Planck Projektgruppe Recht der Gemeinschaftsgüter 1998/3, Max-Planck-Projektgruppe, Bonn.

Heritier, A., Knill, C. and Mingers, S. (1996), *Ringing the Changes in Europe: Regulatory Competition and Redefinition of the State*, Britain, France, Germany, Walter de Gruyter, Berlin.

Hey, C. (2000), 'Zukunftsfähigkeit und Komplexität: Institutionelle Innovationen in der EU', in von Prittwitz, V. (ed.) *Institutionelle Arrangements in der Umweltpolitik: Zukunftsfähigkeit durch innovative Verfahrenskombinationen?*, Leske und Buderich, Opladen, pp.87-100.

Holzinger, K. (1994), *Politik des kleinsten gemeinsamen Nenners? Umweltpolitische Entscheidungsprozesse in der EG am Beispiel der Einführung des Katalysatorautos*, Sigma, Berlin.

Holzinger, K. (1999), *Optimal Regulatory Units: A Concept of Regional Differentiation of Environmental Standards in the European Union*, Preprint from the Max-Planck Projektgruppe Recht der Gemeinschaftsgüter 1999/11, Max-Planck-Projektgruppe, Bonn.

Homeyer, I.V. and Carius, A. (2000), 'Die Osterweiterung der Europäischen Union als Herausforderung für die Umweltpolitik', *Zeitschrift für Umweltpolitik und Umweltrecht*, 3/ 2000, pp.337-368.

Homeyer, I.V., Kempmann, L. and Klasing, A. (1999), 'EU Enlargement: Screening Results in the Environmental Sector', *Environmental Law Network International (ELNI) Newsletter* 2/99, pp.43-47.

Homeyer, I.V., Carius, A. and Bär, S. (2000), 'Flexibility or Renationalization: Effects of Enlargement on EC Environmental Policy', in Cowles, M. G. and Smith, M. (eds), *The State of the European Union. Risks, Reform, Resistance and Revival*, Vol. 5, Oxford University Press, Oxford, pp. 347-368.

Hooghe, L. (1996), *Cohesion Policy and European Integration: Building Multilevel Governance*, Oxford University Press, Oxford.

Immergut, E. (1992), *Health Politics. Interests and Institutions in Western Europe*, Cambridge University Press, Cambridge.

Jacoby, W. (1999), 'The Reality Behind the Potemkin Harmonization. Priest and Penitent: The European Union as a Force in the Domestic Politics of Eastern Europe', *East European Constitutional Review*, http://www.law.nyu.edu/eecr/vol8num1-2/special/ priestpen.html.

Jordan, A. (1999), 'Editorial Introduction: the Construction of a Multilevel Environmental Governance System', *Environment and Planning C: Government and Policy*, Vol. 17, No. 1, pp.1-17.

Jordan, A. (1999a), 'The Implementation of EU Environmental Policy: A Policy Problem Without a Political Solution?', *Environment and Planning C: Government and Policy*, 17, No. 1, pp.69-90.

Jordan, A. (1999b), 'European Community Water Policy Standards: Locked in or Watered Down?', *Journal of Common Market Studies*, Vol. 37, No. 1, pp.13-37.

Jordan, A. and Lenschow, A. (2000), ''Greening' the European Union: What Can be Learned from the 'Leaders' of EU Environmental Policy', *European Environment*, Vol. 10.

Jordan, A., Brouwer, R. and Noble, E. (1999), 'Innovative and Responsive? A Longitudinal Analysis of the Speed of EU Environmental Policy-Making 1967-97', *Journal of European Public Policy*, Vol. 6, No. 3, pp.376-398.

Knill, C. and Lenschow, A. (1997), 'Coping with Europe: The Impact of British and German Administrations on the Implementation of EU Environmental Policy', *EUI Working Paper*, RSC No. 97/57, European University Institute, Florence.

Knill, C. and Lenschow, A. (1999), 'Neue Konzepte – alte Probleme? Die institutionellen Grenzen effektiver Implementation', *Politische Vierteljahresschrift*, Vol. 40, No. 4, pp.591-617.

Knill, C. and Lenschow, A. (2000), *Seek and Ye Shall Find! Linking Different Perspectives on Institutional Change*, Preprints from the Max-Planck-Projektgruppe Recht der Gemeinschaftsgüter, 2000/6, Max-Planck-Projektgruppe, Bonn.

Kraemer, R.A. and Mazurek, A.-G. (1999), *Koordinationsstrukturen zur nachhaltigen Entwicklung in den EU Mitgliedstaaten*, Unpublished study for the Austrian Ministry of the Environment, Ecologic, Berlin.

Liberatore, A. (1991), 'Problems of Transnational Policymaking: Environmental Policy in the European Community', *European Journal of Political Research*, Vol. 19, No. 2&3, pp.281-305.

Liefferink, D. and Andersen, M.S. (1998), 'Strategies of the 'Green' Member States in EU Environmental Policy-Making', *Journal of European Public Policy*, Vol. 5, No. 2, pp.254-270.

Majone, G. (1994), *The European Community as a Regulatory State*, Lectures given at the Academy of European Law, July 1994, European University Institute, Florence.

Marks, G. (1993), 'Structural Policy and Multilevel Governance in the EC', in Cafruny, A. and Rosenthal, G. (eds), *The State of the European Community. The Maastricht Debates and Beyond*, Vol. 2, Lynne Rienner, London, pp.391-410.

Matthews, D. (1999), 'The Ebb and Flow of EC Environmental Instruments: Why the Need for a New Framework Approach to Community Water Policy?' Paper presented at the Sixth ECSA Biennial International Conference, 2-5 June 1999, Pittsburgh.

Mayntz, R. and Scharpf, F.W. (1995), 'Der Ansatz des akteurszentrierten Institutionalismus', in Mayntz, R. and Scharpf, F.W. (eds), *Gesellschaftliche Selbstregelung und politische Steuerung*, Campus, Frankfurt/Main, pp.39-72.

Mazey, S. and Richardson, J. (1993), 'Environmental Groups and the EC: Challenges and Opportunities', in Judge, D. (ed.), *A Green Dimension for the European Community: Political Issues and Processes*, Frank Cass, London, pp.109-128.

McCormick, J. (1994), 'Environmental Policy: Deepen or Widen?', in Laurent, P.-H. and Maresceau, M. (eds), *The State of the Union. Deepening and Widening*, Vol. 4, Lynne Rienner, London, pp.191-206.

Nicolaides, P. (1999), *Enlargement of the EU and Effective Implementation of Community Rules: An Integration-Based Approach*, EIPA Working Paper 99/W/04, European Institute for Public Administration, Maastricht.

Office of Technology Assessment (1994), *Industry, Technology, and the Environment: Competitive Challenges and Business Opportunities*, OTA-ITE-586, OTA, Washington, D.C.

Organisation for Economic Co-operation and Development (1999), *Environment in the Transition to a Market Economy. Progress in Central and Eastern Europe and the New Independent States*, Centre for Co-operation with Non-Members, OECD, Paris.

Peters, B.G. (1992), 'Bureaucratic Politics and the Institutions of the European Community', in Sbragia, A. (ed.), *Euro-Politics. Institutions and Policymaking in the 'New' European Community*, The Brookings Institution, Washington, D.C., pp.75-122.

Peterson, J. (1995), 'Decision-Making in the European Union: Towards a Framework for Analysis', *Journal of European Public Policy*, Vol. 2, No. 1, pp.69-93.

Pollack, M.A. (1999), 'Delegation, Agency and Agenda Setting in the Treaty of Amsterdam', *European Integration Online Papers*, EioP, 3 (6), http://eiop.ot.at/eiop/texte/1999-006a.htm.

Prittwitz, V.V. (1990), *Das Katastrophenparadox. Elemente einer Theorie der Umweltpolitik*, Leske und Buderich, Opladen.

Regional Environmental Centre (1994), *Strategic Environmental Issues in Central and Eastern Europe*, Regional Report 1, Regional Environmental Center, Budapest.

Richardson, J. (1996), 'Eroding EU Policies: Implementation Gaps, Cheating and Re-steering' in Richardson, J. (ed.), *European Union: Power and Policy-Making*, Routledge, London, pp.278-294.

Sbragia, A. (1996), 'Environmental Policy: The Push-Pull of Policy-Making', in Wallace, H. and Wallace, W. (eds), *Policy-Making in the European Union*, 3rd edition, Oxford University Press, Oxford, pp.235-255.

Scharpf, F.W. (1988), 'The Joint Decision Trap: Lessons from German Federalism and European Integration', *Public Administration*, Vol. 66, No. 3, pp.239-278.

Scharpf, F.W. (1994), 'Community and Autonomy: Multi-Level Policy-Making in the European Union', *Journal of European Public Policy*, Vol. 1, No. 2, pp.219-242.

Scharpf, F.W. (1996), *Problem-Solving Capacity of Multi-Level Governance Structures*, Paper presented at the Conference Social Regulation Trough Committees: Empirical Research, Institutional Politics, Theoretical Concepts and Legal Developments, 9 and 10 December 1996, Department of Law, European University Institute, Florence.

Scharpf, F.W. (1997), *Globalisierung als Beschränkung der Handlungsmöglichkeiten nationalstaatlicher Politik*, MPIfG Discussion Paper 97/1, Max-Planck-Institut für Gesellschaftsforschung, Köln.

Scharpf, F.W. (1998), 'Die Problemlösungsfähigkeit der Mehrebenenpolitik in Europa', in Kohler-Koch, B. (ed.), *Regieren in entgrenzten Räumen*, PVS Sonderheft 29, pp.121-144.

Scott, J. (2000), 'Flexibility in the Implementation of EC Environmental Law', in Somsen, H. (ed.) *Yearbook of European Environmental Law*, Vol. 1, Oxford University Press, Oxford, pp.37-60.

Soil and Water Ltd. (2000), *Development of Synthesis Reports for Approximation of EU Environmental Legislation*, Final Report, MC-112, Development of Implementation Strategies for Approximation in Environment, Vantaa, Soil and Water Ltd.

Transport and Environment, European Federation for Transport and Environment (2000), 'Special Feature: Accession States – Golden Opportunity to Size or Miss', *Transport and Environment Bulletin*, European Federation for Transport and Environment, Brussels.

Weizsäcker, E.U.V. (1990), *Erdpolitik. Ökologische Realpolitik an der Schwelle zum Jahrhundert der Umwelt*, Wissenschaftliche Buchgesellschaft, Darmstadt.

Zito, A.R. (1999), 'Task Expansion: A Theoretical Overview', *Environment and Planning: Government and Policy*, Vol. 17, No. 1, pp.19-35.

5 Analytical Frameworks for Policy and Project Evaluation: Contextualising Welfare Economics, Public Choice and Management Approaches

WAYNE PARSONS

The Challenge of Integrating Knowledge

The issue of evaluation in public policy is a complex and deeply problematic area of both theory and practice. In this chapter we take a very broad view of evaluation to include policy analysis (as a field) as well as an aspect of policy planning. In simple terms evaluation is about 'what works', when, and for whom. Evaluation is the activity through which policy makers and those involved in policy-driven research endeavour to get information and knowledge out of problems so as to better understand the ways in which policies, programmes, and projects have brought about, or failed to bring about change, or how a proposed intervention is likely to impact on a given problem. How this task of accessing knowledge and using evidence may be approached, of course, depends on what kind of analytical frames are being used by policy makers and policy analysts.

Evaluation is fundamentally a process of *valuing*, and different frames of value inevitably generate different ways of thinking about the problem of finding out what works, when, how, and for whom. Evaluation, therefore, directs our attention to the issue of (what one of the founding fathers of the policy sciences, Harold Lasswell, saw as) the task of integrating knowledge so as to contribute to the democratisation of humankind (Lasswell, 1970). In Lasswell's sense evaluation involves 'who gets what knowledge, when and how?' (Lasswell, 1958).

The central question to ask about evaluation, therefore, is: whose *values* get to dominate? As Francis Terry notes: 'a consideration of "what works" in transport policy is at least partially determined by your philosophical starting point and the key questions are easier to answer at the tactical level than to do so in terms of total strategy' (Terry, 2000, p.192). When we are discussing competing ways of obtaining knowledge about policy interventions in public problems, we are focusing on a major issue of democratic governance: the *production* and *use* of knowledge about social, economic, environmental and other problems. The task of *integrating* evaluation research for the purpose of improving policy making, however, must not to be seen as an exercise in blurring the edges, so much as making the edges clearer. *Integrating knowledge is fundamentally to do with designing methods, processes and institutions which can best serve to clarify competing arguments, 'stories', forecasts, discourses, and knowledge claims.*

The task of the policy sciences, Lasswell believed, was to facilitate the integration of knowledge about human problems: the great challenge for democracy was to ensure that different forms of knowledge, and sets of values could be used to develop more communicative and knowledgeable forms of governance. However, given the fact that there are deep and profound disagreements and differences of opinion and values, the idea of 'integrating' what we know about evaluation seems a contradiction in terms. To what extent, for example, is it possible or indeed desirable, to 'integrate' different and apparently incommensurate analytical frameworks? Does integration involve the idea of one 'big' 'general theory' of evaluation? Does evaluation actually *require* the development of some kind of Esperanto, or common language in order to become more useful for policy makers? Clearly the answer to this must be an emphatic *no.* Different approaches to evaluation embody, as it were, different 'assumptive worlds' (Young, 1977) or 'frames' (Schön and Rein, 1994). When Lasswell stressed the importance of *integration* it was not to promote a kind of intellectual unity or theory of everything, but quite the opposite: integration for Lasswell involved a process of *clarification.* The aim of the policy sciences was to promote a greater clarification of the values which knowledge *in, for* and *of* the policy process embodies. In seeking to integrate knowledge of evaluation *in, for* and *of* the policy process, therefore, the aim should likewise be to clarify different ways of thinking so as to understand what approaches work best in what contexts, when, how and for whom.

This chapter argues that if we are to better integrate our knowledge of evaluation we should proceed by recognising that our knowledge is located in a variety of different and competing frames. This idea of 'frames' is drawn from Donald Schön and Martin Rein's work on 'frame-reflective discourse'. Policy frames, they argue, are: 'the taken-for-granted assumptional structures of policy research' (Schön and Rein, 1994).

> We see policy positions as resting on underlying structures of belief, perception, and appreciation, which we call 'frames'. We see policy controversies as disputes in which the contending parties hold conflicting frames. Such disputes are resistant to resolution by appeal to facts or reasoned argumentation because the parties' conflicting frames determine what counts as fact and what arguments are taken as relevant and compelling. Moreover, the frames that shape policy positions and underlie controversy are usually tacit, which means that they are exempt from conscious attention and reasoning (Schön and Rein, 1994, p.23).

This focus on frames marks a radical departure from policy analysis predicated on the ideas of rational decision making, instrumental rationality and value neutrality which have for so long dominated the field. For Schön and Rein the resolution of frame conflict involves the development of a reflective policy conversation.

> Participants in such a conversation must be able to put themselves in the shoes of other actors in the environment, and they must have a complementary ability to consider how their own action frames may contribute to the problematic situations in which they find themselves. If policy makers are to communicate reliably with their antagonists and reliably interpret flaws in the design of policy objects, they must be able to give 'good reason' to other actors in the environment, which means entering into the action frames that inform multiple constructs of reality. The very act of giving reason means that policy makers must also be able to reflect on the action frames that underlie the transactions through which they may have helped to promote miscommunication or exacerbate design flaw. Similarly, in order to contribute to the re-framing of policy dilemmas, policy makers must be able to reflect on the action frames held by their antagonists. Even to recognize the existence of such dilemmas, policy makers must be able to reflect on their own action frames: they must overcome the blindness induced by their own ways of framing the policy situation in order to see that multiple policy frames represent a nexus of legitimate values in conflict (Schön and Rein, 1994, p.187).

The aim of this chapter is to map out and reflect on the relationship between different analytical frames and their assumptive worlds or contexts. Integration, *per se*, can only come about in Schön and Rein's sense through reflection in practice. That is to say, in *specific* contexts and problem situations in which an evaluative conversation takes place. In designing evaluative strategies, therefore, attention should be paid to facilitating this conversational/reflective process between frames. The integration process, from this point of view, should be about advancing the development of a complementary understanding of how proponents of competing evaluative frames can 'talk' to one another. Figure 5.1 is designed as an aid to promote such a conversation through 'mapping' some of the main evaluative frames.

Contextualising Evaluation Frames

For the purpose of this paper we can identify eight frames for policy and project evaluation. Although we have set out eight frames, it is important to note from the outset that this by no means underestimates the significance of the differences *within* the various frames and how approaches can cut across each frame. So called 'evidence based' approaches (see Davies, Nutley and Smith, 2000), for example, could be situated in more than one of the 'mainstream' and 'alternative' frames. A more comprehensive account than is offered in this chapter would, most likely, set out not eight but eighty-eight frames. Each one of these frames is composed of a range of different approaches that to capture them in one 'frame' is quite impossible. However, in order to make it easier to initiate a discussion of integrative strategies for evaluation, we here simplify to broad statements of commonality.

Figure 5.1   Map of Evaluative Frames

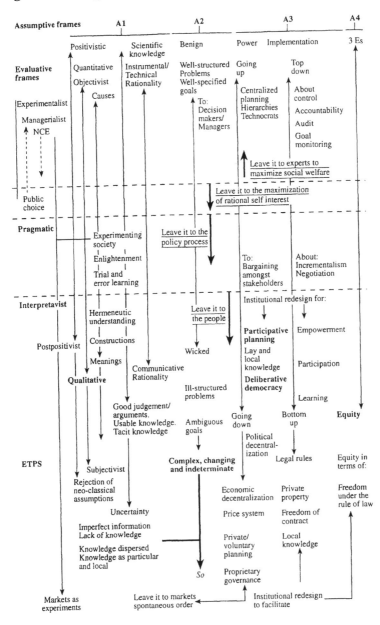

Most of these frames are accepted 'mainstream' modes of evaluation. However, we have also included approaches which are somewhat more 'critical', 'radical' or 'non-mainstream' for an area like transport which has long been dominated by evaluative methods framed by neo-classical economics and its variants. In particular we focus on 'interpretivist', critical realist and what we can term 'Evaluation through the Price System' (ETPS).

The aim here is to provide a broader setting in which to locate the 'mainstream' frames deployed by transport evaluation. In doing so, we want to suggest that the 'critical' frames can be understood as essentially *complementary* to the mainstream techniques of transport evaluation. And, that in designing evaluative strategies for transport policies and projects, more account should be taken of the implications of the critical frames.

The 'mainstream' frames comprise five types:

- *E1. Neo-Classical Economics (NCE)*: Evaluation as calculating and distributing cost and maximising benefits and social welfare.
- *E2. Experimentalism*: Evaluation which seeks to approximate to natural scientific methods. This would also include forecasting/predicting such issues as traffic growth.
- *E3. Managerialism*: Evaluation as the measurement and monitoring of performance.
- *E4. Public Choice*: Evaluation as facilitating the maximisation of rational self-interest.
- *E5. Pragmatism*: Evaluation as the enlightenment of decision makers and stakeholders.

Clearly there are significant differences between these approaches which should not be overlooked, but for the purpose of this chapter I want to suggest that an integrative strategy should pay more attention to the role which alternative, non-mainstream approaches can play in helping us to understand the position of 'mainstream' evaluation strategies. There are three such non-mainstream approaches I would like to discuss here:

- *E6. Interpretivism*: Evaluation as education and empowerment of all stakeholders through dialogue and deliberative democracy.
- *E7. Evaluation through the Price System (EPTS)*: Evaluation as allowing markets to facilitate experimentalism, learning and self-organisation in conditions of uncertainty and complexity.

- *E8. Critical Realism*: Evaluation as the process of identification, articulation, testing and refinement of CMO (context – mechanisms – outcome) configurations.

Interpretivism

This frame has developed as an alternative response and critique of mainstream rational policy analysis. It rejects the positivist and objectivist claims of the other frames in favour of a hermeneutic, constructivist or argumentative approach. Its influences range from ethnography, phenomenology and French and German critical theory. Interpretivists argue the case for policy analysis and/or evaluation as an empowering learning process rather than as a technocratic, managerialist exercise (see, for example, Fischer, 1995; Guba and Lincoln, 1990).

An instance of this approach deployed in the field of transport was the use of interpretive method[1] in 1998 as part of the drafting process of the Dutch National Transportation Plan. This involved over 80 participants from local, regional, provincial and national governments together with a variety of stakeholder representatives.[2] The aim of the research was to make better sense of the conflicting arguments that frame and drive Dutch transport policy. Eventually four key arguments were identified. These were framed around, respectively, issues of scarcity, logistics, pragmatism and technology. The research clarified the kind of policy implications which followed from different frames, and the kind of management regimes which these frames implied. By clarifying the different frames and assumptions underpinning arguments which are central to debates on Dutch transportation policy the research hopes to make the intractable issues surrounding transport choices more 'tractable' and thereby improve policy deliberation and learning. Interpretive analysis of this kind does not aim to replace mainstream evaluation approaches, but seeks to ensure that their values, assumptions, and beliefs and the institutional, and organisational implications of their positions are contextualised and understood.

Evaluation through the Price System

This is a frame not generally seen in terms of 'evaluation' *per se*. Hayek and his followers, however, would argue that as the world is an uncertain, complex, unknowable place any attempt to try and 'evaluate' or forecast and predict in a comprehensive way the present and future demand and supply of transport is doomed to fail just like any attempt to plan the economy (Hayek, 1944, 1948, 1952, 1957). The price system in a Hayekian sense is a form of *hermeneutic* evaluation in that it is markets which are considered to be the best (indeed the only) way to *interpret* the mass of information and knowledge in a complex and uncertain world. Daniel Klein gives an idea as to what this would mean in the case of a Hayekian approach to urban transit. Out would go city-wide transport evaluation studies, and in would come private property and freedom of contract.

> This means privatization of the sidewalks, shelters, streets and highways. Private ownership of the streets would mean the existence of a party that has the ability and motivation to deal with information problems, passenger facilities, network effects, disjointed pieces, destructive competition, interlopers and curb-side conflicts (...) Imagine the city streets and roads divided up into segments or small districts. Each separate unit would be under the control and management of a private entity. We can begin to think of the entire city as a continuous patchwork of shopping-mall roads and drive-ups. Just as shopping malls allow free parking, street owners might make road access one of gratis attractions to visitors, residents and businesses. Just as proprietary communities often provide minibus service gratis, the road-owner might provide a free bus service. Alternatively, the road-owners might implement electronic road pricing (Klein, 1998, pp.6-8).

In this brave new world (proprietary) governance would mean services being broken down into 'independent offerings based on smaller geographical units' (Klein, 1998, p.6). Evaluation, forecasting and predicting, of course, would still have a place but it would be small, particular, localised and be sensitive to the way in which markets valued given transport arrangements. Evaluation would be what markets do.

> The plan of government action that would best induce the utilization of local knowledge is a plan to privatize the urban landscape and to enforce the contracts of private parties (...) The invisible hand is without central planning, but not without planning. It moves by a web of voluntary planning. Voluntary planning is usually better informed, more responsive, more intelligent, and more humane than government planning (Klein, 1998, p.9).

In the same way, ETPS would not mean abandoning evaluation, so much as 'centralised' evaluation. It would move through a 'web' of voluntary evaluation, and consequently be 'more informed, more responsive more intelligent and more humane' (Klein, 1998, p.9). The government's role would be to ensure that effective use is made of *local knowledge*, to facilitate (voluntary) transport planning through the workings of private property and contracts. For the ETPS frame, the issue is not whether to have evaluation or not, but whether it is to be done centrally or in a decentralised form so as to make effective use of 'particular knowledge of local circumstances' (Hayek, 1960, p.352).

Critical Realism

This is an approach based on the philosophy of critical realism as developed by Bhaskar (1989, 1997) and others.[3] This frame has not been included on figure 5.1 as it is rather problematic to place in this diagram. It belongs alongside E6 and E7 even though it is at odds with their subjectivism and, in the case of interpretivism, constructivism. The frame is useful for the purposes of thinking about integrative strategies as it seeks to chart a middle path between the experimentalism, positivism and pragmatism of E1, 2, 3, and 5 and the methodological individualism of E4, on the one hand, and the constructivism of E6.

Ray Pawson and Nick Tilley have advanced a persuasive case for an approach to evaluation which is based on critical realism. As they point out, in its early days evaluation was framed by positivistic experimentalist ideas, which in turn gave rise to a more pragmatic approach and in more recent years constructivist and pluralist modes of evaluation. They come, however, not to bury scientific evaluation, but to rescue it from positivism: 'We are indeed', they proclaim, 'the courtiers at the palace of scientific evaluation, announcing that the king is dead – long live the king!' (Pawson and Tilley, 1997, p.30). This involves a shift from a 'successionist' to a 'generative' theory of causation. Experimental methods of evaluation, they claim, have proved so disappointing because of its very logic which 'either ignores (...) underlying processes, or treats them incorrectly as inputs, outputs or confounding variables, or deals with them is a *post hoc* and thus arbitrary fashion' (Pawson and Tilley, 1997, p.54). The logic of realist explanation, however, suggests that:

> The basic task of social inquiry is to explain interesting, puzzling, socially significant regularities (R). Explanation takes the form of positing some underlying mechanism (M) which generates the regularity and thus consists of propositions about how the interplay between structure and agency has constituted the regularity. Within realist investigation there is also

investigation of how the workings of such mechanisms are contingent and conditional, and thus only fired in particular local, historical or institutional contexts (C) (Pawson and Tilley, 1997, p.71).

Programmes 'work', they argue, through their ability to 'break into the existing chains of resources and reasoning which led to the problem'. The aim of evaluation should be to focus on the questions of 'what are the mechanisms for change triggered by a program and how do they counteract the existing social processes?' (Pawson and Tilley, 1997, p.75); and 'what are the social and cultural conditions necessary for change mechanisms to operate and how are they distributed within and between contexts?' (Pawson and Tilley, 1997, p.77). Explanations should, they suggest, be structured around three factors: context (C), mechanism (M) and outcome (O). The task of realist evaluation is to find ways of 'identifying, articulating, testing and refining conjectured *CMO* configurations' (Pawson and Tilley, 1997, p.77). Given this, realistic evaluation research is '*CMO*' rather than programme driven. Realist analysis reveals that programmes are not 'unitary happenings' that work or don't work: the sheer variation in context gives rise to considerable complexity in how a given policy intervention will impact upon a problem. Instead of asking the (rather silly) question about 'how come a programme works/doesn't work?', realists posit a far more scientific (and rather Lasswellian) question regarding evaluation: 'what is it about a given policy, programme or project that works, for whom, where, when and how?'.

Contextualising Assumptive Frames

As frames these various approaches to getting knowledge out of policy interventions (or accessing knowledge) embody different 'assumptive worlds'. That is, they constitute competing, conflicting, and often incommensurate mind-sets. Evaluative frames provide different ways of seeing, and suggest different ways of doing, and different ways of valuing. They embody a considerable range of ideas about social and individual welfare, the role of the state, market, and community, as well as the policy process, and problems of implementation, and ideas about democracy.

We can identify four key *assumptive frames* which may serve to show how these different *evaluative frames* stand in relation to one another (see figure 5.1). They comprise:

- A1: Assumptions about methodology, ontology and epistemology (MOE assumptions).
- A2: Assumptions about the complexity of problems.
- A3: Assumptions regarding decisional contexts, institutional settings and policy processes.
- A4: Assumptions about welfare.

Assumptions About Methodology, Ontology and Epistemology (MOE)

What kind of assumptions does a given frame have about issues such as the positivistic, quantitative nature of problems? How does it view the nature of 'reality' and the kind of knowledge which is or can be produced? What kind of methodology does a frame favour?

Our mainstream approaches tend towards a MOE profile which is positivistic, objectivist, causal, concerned with acquiring 'scientific' type knowledge about evaluation problems. In broad terms one could argue that for the (positivistic) mainstream reality is seen to exist, 'out there', and we can get to know the laws of cause and effect which structure the real world; it can be measured, and studied in a value-free way; and we can test our theories and hypotheses in an empirical manner. In its purest form, of course, this would approximate to classical experimental evaluation which seeks to emulate the experimental method of the natural sciences. Neo-classical economics with its roots in physics envy and Newtonian mechanics is very much in the positivistic domain as is public choice theory. Both share a core assumption about the nature of human behaviour, and the possibilities of framing a science around the simple and crude assumption that *homo economicus* is (*ceteris paribus*) a self-interested maximiser. 'New Public Management' is also deeply positivistic in its belief that it is possible to measure and calculate, and define clear goals and objectives.

Pragmatic approaches take a different direction in viewing the issue of evaluative method and knowledge in ways which emphasise the so-called 'enlightenment' role of evaluation and Popperian experimentation. Pragmatism, thus, is a 'middle way' seeing reality as something which can be known in practice and is mediated through bargaining and negotiation. Also establishing a middle way albeit of a different kind is the E8 critical realist frame: it believes in a science of society, and the possibilities of objectivity and detachment, but rejects the ontology, epistemology and methodology of classical experimental evaluation. As Harvey and Reed argue:

> From critical realism's perspective, there is the real world and what we say about that world. Contrary to idealist epistemologies, our statements about the structure of the world are not the world as such, and the two must never be confused. By the same token, the empirical appearance of reality should never be mistaken for those generative structures and mechanisms whose powers and limitations actually produce the world of appearance. Whilst experience can often be the point of departure for scientific inquiry, empirical presence alone cannot give us access to the lawful structures that produce those appearances. To penetrate to that more fundamental substratum of causal forces, more than controlled observation is required. Just as Kant rejected Hume's empiricism, so critical realism rejects the claim that the experimental method, in and of itself, can provide logically feasible laws. Causal explanations derived from empirical evidence are necessary, but never sufficient, for the identification of law-like principles. The same assertion holds equally for social scientific surveys, demographic analyses, or participant observer studies. The empirical world can reflect, but never give us direct access to the causal mechanisms of reality (Harvey and Reed, 1996, p.301).

At the opposite end of the scale to frames E1,2,3,and 4, interpretivism and ETPS (Frames E6 and 7) are postpositivist, subjectivist and hermeneutically orientated. Hayek, of course, utterly rejects the kind of MOE assumptions that underpin the arguments of neo-classical economics and public choice frames. He sees the social world in subjectivist terms as an unknowable, uncertain place and is dismissive of attempts to apply the methods of the natural sciences to the social order. Interpretivist approaches, of course, cover a good deal of ground, encompassing the MOE assumptions of post-positivism, post-modernism and critical theory. For interpretivists, objective, value free knowledge is not possible in evaluation research or in any other kind of social research. Whereas frames E1-5 would tend, therefore, to emphasise the measurability of policy success/failure, the pragmatic and critical frames would question how measurable policy impacts and outcomes can be. In terms of sustainability, for example, frames would make different assumptions or claims about the 'measurability' of 'sustainable transport'.

Assumptions about the Complexity of Problems

How does a given frame see policy problems: as 'benign' and well-structured, or 'wicked', complex and ill-structured? These are a crucial set of assumptions for understanding evaluation frames.

The mathematician and designer Horst Rittel argued that we confront a world of problems which are infinitely more confusing than is generally recognised by policy designers and planners. Rittel (Rittel and Webber, 1973) distinguishes between 'tame' or 'benign' and 'wicked' problems.

The problems that scientists and engineers have usually focused upon are mostly 'tame' or 'benign' ones. As an example, consider a problem of mathematics, such as solving an equation; or the task of an organic chemist in analysing the structure of some unknown compound; or that of the chess player attempting to accomplish checkmate in five moves. For each the mission is clear. It is clear, in turn, whether or not the problems have been solved. 'Wicked' problems, in contrast, have neither of these clarifying traits; and they include all public policy issues – whether the question concerns the location of the freeway, the adjustment of a tax rate, the modification of school curricula, or the confrontation of crime (Rittel and Webber, 1973, p.160).

Policy problems are 'malignant', 'vicious circles', 'tricky', and 'aggressive' and it is very dangerous for us to treat them as if they were 'tame' and 'benign'. Because of the 'wicked' nature of public policy issues, the attempt to use rational analytical techniques to understand them, and 'solve' them is fundamentally misguided. 'Wicked' problems exhibit tendencies towards, for example:

- No definite formulation of what the problem is. Formulations of wicked problems oftentimes correspond to their solution.
- Not having a stopping rule.
- Not having a test of what constitutes a solution.
- Not having innumerable sets of potential solutions or admissible operations.
- Being the symptom of another problem.
- Having many possible explanations.
- Being unique (Rittel and Webber, 1973, pp.161-66).

To use the language of Herbert Simon (see Parsons, 1995, pp.354-5), 'wicked' problems are essentially *ill-structured* as opposed to being *well-*

structured problems. 'Wicked' problems are messy, ill-defined, indeterminate conditions, and the more wicked they become, the less are they amenable to the techniques of rational analysis. From the point of view of evaluation one could argue that it is well-structured problems which are suitable cases for instrumental or technical rationality. The NCE evaluative frame, for instance, could be seen as most appropriate when we confront well-structured or 'tame' problems in which the costs and benefits are known and can be calculated.

The design of evaluation strategies has, therefore, to take account of the different problem settings within which decision making takes place, that is to say, the degree and kind of problem *complexity* facing decision-makers. This is particularly important in the context of evaluating *policies* as opposed to plans and programmes and as between the appraisal of specific projects and the assessment of the more strategic levels of policy design. As one moves from the domain of policy design to the design of specific programmes and projects one could argue that uncertainty begins to fall, and rational techniques become more appropriate to the evaluation of given processes or outcomes (see figure 5.2.).

At policy level, the relationship between different policy objectives and programmes becomes inevitably more indeterminate and uncertain. Consensus about values, valued outcomes, and agreement about aims and objectives are clearly much higher at the project level, than the policy level, and the temporal context (long-term vs. short-term) is also of considerable significance. From the perspective of the ETPS frame, it could be argued that complexity is best handled through markets: the more complex the problem, the more do markets reveal themselves as better able to ensure the delivery of the '3Es': economy, efficiency and effectiveness. The more complexity there is, the less desirable is it to leave evaluation in the hands of bureaucrats and politicians (supposedly) seeking to maximise and calculate the public interest.

In terms of the power context, the more we move towards post-positivist methods, the more does evaluation become concerned with politics, empowering, participation, democratisation and marketisation. This is to say, the greater the 'wickedness' of problems, the more appropriate are methods which seek to acknowledge the political, constructivist, and hermeneutic tasks of evaluation in democratic societies. From the point of view of the positivistic frames, problems are viewed as occupying the 'benign', well-structured part of figure 5.1. This is where 'we want to be': in the world of facts. Alas, of course, this is where we rarely, if ever, are. Decision makers in particular want to keep problems in the realm of the 'benign' and 'tame' because it justifies and legitimates their approach to finding out what works or not: as a matter of fact. Framing problems as well-structured facilitates an up-wards distribution of power. Decision makers define (determine) the problems, the agenda, the solution and the policy or programme discourse. The further evaluation shifts in the direction of problem wickedness, however, the more indeterminate things become, and the more does evaluation become enmeshed in value conflicts and contested meaning: that is to say, politics.

Figure 5.2 Project versus Policy Evaluation

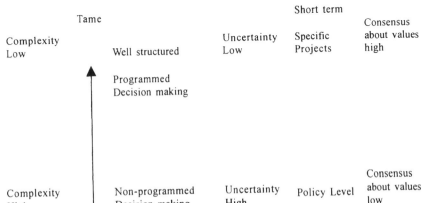

Assumptions regarding Decisional Contexts, Institutional Settings and Policy Processes

How does a given evaluative frame situate evaluation in the context of power, organisational and institutional forms (hierarchy, market, community) and implementation strategies? How is a given evaluation strategy located in the context of the policy process? How is, for example, a strategy being deployed by decision makers or power holders? Who is empowered and who is dis-empowered in a given strategy? How does an evaluation strategy relate to the implementation design? Evaluation has to be understood in the context of specific policy processes. In very simple terms this involves understanding how evaluation (*ex-ante* and *ex-post*) is *used* to define problems, frame solutions and shape decisions. That is, how it is utilised by all stakeholders – including policy makers and street-level (project or programme) implementators.

Such assumptions involve the idea that all information and policy relevant knowledge is embedded in institutional and organisational contexts. Information has, as it were, a 'social life' (Brown and Duguid, 2000). When we are considering the process by which values are obtained, by whom, when and how, for example, we are acknowledging that evaluation is quintessentially about organisational and political empowerment and dis-empowerment. One key question, therefore, to ask about any evaluation frame is: who or what is empowered by this way of thinking? That is to say, whose values, and what values win, and whose/what values lose? *Qui bono?* And, when and how do they win and lose? Evaluation strategies are developed and are implemented in the context of institutions and power relationships. The task of integrating different approaches cannot progress very far if power – who gets what, when and how – is ignored.

Indeed, one could tell the story of evaluation in terms of how different frames distribute or allocate power. The choice of evaluative method is rooted in the exercise of power and discipline. In the case of economic approaches, for instance, such as those deriving from neo-classical economics and more recently public choice, one could argue that it has tended to empower those deploying economic discourse and instrumental rationality (cost-benefit analysis and its variants) and those deploying the language of rational self-interest. And, as a consequence, it disadvantages those who do not, cannot or will not see the problem of evaluation in the terms set out by economistic evaluative frames, or whose arguments cannot be expressed in the same terms: they might be to do with aesthetic or ethical arguments rather than those couched in the language of efficiency or effectiveness. Managerialist approaches to evaluation, on the other hand, serve to *em*-power those who exercise managerial control, and *dis*-empower those who are on the receiving end of audit and performance management techniques. Those who advocate interpretivist, or constructivist frames, of course, argue strongly for evaluation being about empowerment, and communicative rationality rather than control and instrumental rationality.

The choice of evaluative frame, therefore, has to be understood in terms of the decisional or institutional settings in which choice arises. One could argue, for instance, that neo-classical economics and management frames are appropriate to decision making and implementation which is concerned with ensuring power stays up, rather than moves downwards.

Decision making may well be driven by the desire to empower technocratic or bureaucratic actors, and to facilitate top-down control in the implementation process. In which case, whosoever pays the evaluator, calls the tune. *Evaluation is not simply about finding out what works, or what can work, it is also about seeking to exercise control within the policy making process, and the implementation and organisational contexts of policy and programmes.* The more concerned policy makers are to deploy evaluation as part of a *control* strategy, the more likely they are to see evaluation in terms of well-structured problems, with well-specified goals. Hence, the political appeal of neo-classical economics, experimentalist, managerialist and public choice frames for policy makers.

Another way to look at the issue of institutional or organisational context is in terms of what Dryzek (1997) has identified as:

• Leaving it to the experts: administrative rationalism.
• Leaving it to the people: democratic pragmatism.
• Leaving it to the market: economic rationalism.

Frames E1, 2 and 3 are clearly are in the domain of 'administrative rationalism': they involve evaluation strategies being determined by 'the experts'. In this context evaluation takes place in organisations and institutions seeking to deploy evaluation as a way of securing hierarchical, technocratic, managerial and/or administrative dominance and control through the use of evaluative processes and output. This may be in the ex-ante or ex-post stages of the policy process: that is, it may involve the use of evaluation to 'set the agenda', define the problems, frame decisions, devise implementation strategies, monitor programmes and projects, and to evaluate impact (in the *Public Interest*). The E4 frame (public choice) would be viewed as 'leaving it to the market', as would E7 (evaluation through the price system). Frame E5 (Pragmatic evaluation) takes place in the context of 'democratic pragmatism': here evaluation is being deployed in the context of a pluralistic process of bargaining, experimentation, trial and error learning and 'enlightenment'. This is the world of incrementalist politics where evaluation is part of the game of politics and policy.

Of course, for Dryzek himself, evaluation has another more radical frame which challenges administrative rationalism, pragmatic politics and self-interest: interpretive or more deliberative forms of evaluation. Frame E6, and E7 are two critical frames that challenge the kind of political, and organisational assumptions of the other frames. The interpretive frame (E6) is about the enhancement of 'voice' through giving greater opportunity for environmental evaluations to be framed by participation, and the shift towards a more deliberative mode of democracy. ETPS (E7) involves another paradigm shift: to use markets as ways of experimenting. To enhance the power of entering and 'exiting' to 'discover new ways of doing things better than they have been done before' (Hayek, 1948a, p.101).

Here it is important to clearly distinguish between two distinct (if related) frames: Public Choice and ETPS (frames E4 and E7). Dryzek, in common with many other advocates of deliberative and collaborative approaches to policy analysis and planning, tends to conflate the two, when they would be far better understood as comprising two very different approaches to public policy. For Hayek, it is the market which can best serve to *co-ordinate* and *interpret* information and knowledge in a complex, uncertain and changeable world. However, in the public choice frame, the key assumption is about rational self-interest as a propensity of human behaviour which is incapable of achieving the 'public interest' as long as it is channelled through bureaucratic and political actors. For public choice, markets make best use of self-interest to secure greater efficiency, economy and effectiveness in the delivery of 'public' as well as 'private' choices. Hayek is not coming from the direction of neo-classical economics, but from an assumptive world which has much in common with the 'deliberative' approach to environmental evaluation favoured by Dryzek. The key point of difference is really over how to interpret information and knowledge in conditions of complexity, when the claim to know is seen to be dubious by both sides. For Dryzek the hermeneutic task is best done through political decentralisation and deliberative democracy. For Hayek it is through economic decentralisation and the price system. For Dryzek policy analysis as a hermeneutic activity can best be secured through activating greater citizen participation, whereas for Hayek it is through private planning and proprietary governance.

Interpretivist and Hayekian frames, therefore, are perhaps far more complementary than they look at first. Indeed, Hayek has far more in common with the MOE assumptions of (post-positivist/post-modern) interpretivist approaches (see Burczak, 1994) than the MOE assumptions (as well as the A2 and A3 assumptions) of the public choice frame or cost-benefit analysis. It may well be that in practice, the call for deliberative modes of democracy and collaborative approaches to planning on the one hand, and evaluation through the price system on the other, are actually far from being incommensurate frames, sharing as they do a concern to access specifically *local* knowledge.

In terms of sustainability, one could argue that such evaluative strategies that may be deployed will reflect the kind of sustainability values or assumptions (or sustainability discourses) which predominate over time and space, and as between different policies, programmes and projects. As the whole concept of sustainability is very flexible and broad-ranging (if not at times downright meaningless) the question arises as to whose definitions win out, when and how? That is, in Dryzek's terms, administratively rational, market rational, or politically rational orientated sustainability? If, for example, sustainability is seen to be a local, bottom-up process, then interpretivist evaluation combined with collaborative planning strategies might be deemed to be most appropriate, in the context of a given policy, programme or project. However, as Mark Pennington observes:

> Seen through a Hayekian lens, collaborative planning overestimates the extent to which social co-ordination can be brought about by deliberative means, a problem that is perhaps reflected in the vague and often contradictory usage of the 'sustainable development' discourse (...) In turn, it is because of these epistemological limits that we may need to rely more on impersonal markets. The constant process of positive and negative feedback embodied in prices, facilitates a process of mutual self-adjustment between people who never actually meet *and cannot* know in sufficient detail the precise circumstances of others. What is still more serious, however, from a Hayekian point of view, is that the institutional arrangements advocated in collaborative planning may actually *reduce* information flows and hence thwart the desired process of inter-subjective social learning. At the core of Hayek's critique of planning is his emphasis on the significance of tacit knowledge. This refers *to time and place specific information that cannot be articulated in verbal form* (Pennington, 2001).

A strategy predicated on sustainability as a state-driven process of modernisation (predicated on the notion that they *do know*) would be more disposed towards neo-classical economics, experimentalist and managerialist frames. However, it is quite possible that evaluative strategies for sustainability could utilise both political and economic decentralisation: hermeneutic evaluation as the enhancement of both exit and voice! Far from exit driving out voice (Hirchman, 1970) increasing exit capacity (in property owning democracies) might also serve to *enhance* voice.[4]

Assumptions about Welfare

To what extent is evaluation located in the context of values such as efficiency, economy and effectiveness (the 3Es) and the maximisation of social welfare, on the one hand, or in terms of equity and ethics, on the other?

Neo-classical economics and managerialist frames are, of course, most preoccupied with the 3Es, and with the calculation of cost and benefit so as to maximise and distribute social welfare. However, as is well-appreciated (Parsons, 1995, p.532), welfare distributions are themselves outcomes of how decision makers choose to weight or balance the relationship between what is fair or equitable and what is efficient. Public choice frames get around this by showing how the mechanism of self-interested maximisation will inevitably fail to deliver efficiency and cost effectiveness as long as bureaucrats and politicians trouble themselves with trying to maximise the public interest through ever bigger budgets and bureaus. Managerialist approaches to evaluation – especially the variety commonly known as 'New Public Management' – have come to emphasise that the 3Es and 'value for money' constitute the primary aims of performance management with its emphasis on target setting, performance indicators, action planning and quantitative measurement.

The hermeneutic evaluation of both interpretivist and Hayekian frames take a very different line on the issue of balancing the demands of efficiency and equity. For the interpretivists, evaluation is about public learning, and as such it is far more concerned with ensuring a more participative and equitable evaluative *process*. The ETPS frame is also far more concerned with process: equity in terms of freedom under the rule of law, and freedom of contract. If the price system is allowed to work, it contends, rather than the evaluations of planners, technocrats and bureaucrats, the market will deliver greater efficiency, effectiveness and economy. Whilst pragmatic pluralist politics will be concerned with equity of outcomes (fairness in access terms), Hayekian liberalism will see the main problem for the state as ensuring a fairness in the due process of law: equality *before the law*. The critical realist frame would also tend to emphasise the importance of evaluation as having a role in wider *learning*, and *emancipation* rather than *control*.

Frames E6,7 and 8, therefore, although coming at the problem of evaluation from *very* different directions, are in the same ball park, even if they are not exactly singing from the same song-sheet. For the critical realists evaluation can be a way in which revised scientific method can facilitate greater understanding of the human condition and policy choices; for the interpretivists evaluation can serve to enhance voice and extend the participation of citizens in the construction of problems and policy options; and for the Hayekian perspective markets are the best way to promote human freedom and experiment with different ways of doing things. In place of the 3Es therefore, critical realism gives us emancipation through a new view of science; interpretivists gives us democracy through deliberation; and ETPS gives us freedom through *market* evaluation.

Integrating Evaluation Approaches

It has undoubtedly been the case that evaluation in the field of transport has tended to be rather closed and less subject to 'contested meaning' than many other areas of public policy. In part this has been due to the fact that transport evaluation has been dominated by experts coming to the field with a background in NEC and engineering. This professional dominance and closure itself has been a consequence of the way in which the issue of evaluating transport policy has been framed by the demand *from decision makers* for knowledge in terms of 'efficiency', 'cost', 'impact', and 'time saving'. In other words, transport evaluation has, for the greater part, taken place in the context of: positivistic 'number crunching', technical rationality, the search for scientific knowledge, and attempts to balance the demands of efficiency and equity, and a relatively 'benign' construction of transport problems. It has been a field framed, in the main, by welfare economics, experimentalist approaches and managerialism usually in the context of a highly pragmatic policy process.

From the perspective of the *critical* frames (E6, 7, and 8) evaluation in the field of transport has been defined by 'leaving it to experts', rather than to 'the people' or to 'markets'. It has generally been framed by a view of the possibilities of social science (qua science) that is wholly rejected by the likes of Hayek, Habermas, or Bhaskar. *The key question to ask, therefore, is what kinds of situations are appropriate for the 'mainstream' frames, and what kind of situations are appropriate for the deployment of the 'critical' frames? In general terms, of course, the answer to this question is: it depends on what values and whose values are winning out.* In the case of sustainability, for example, the evaluation strategy will depend on what construction is put upon the idea of sustainability itself. Radical, 'deep green' or 'strong' accounts of sustainability would be likely to come at the issue of transport from the perspective of evaluation as a bottom-up, 'empowering' process (Dryzek, 1997; Fischer, 1995). Whereas, those advocating an ETPS strategy would argue for 'leaving it to markets': private/voluntary planning rather than planning by government (see, for example, Anderson and Leal, 1991; or Klein, 1998). Sustainability viewed in a more technocratic or manageralist way, on the other hand, would see the problem of evaluation more in terms of using knowledge to manage, plan, and solve problems, meet targets and set environmental goals.

There is no denying that these competing frames constitute very different ways of seeing the issue of transport evaluation in the context of environmental sustainability. What kind of *reflective conversation*, however, is it possible to imagine the proponents of these frames having in the real world? How could such a conversation be structured? If we regard the task of integrating as a process which involves (in Lasswell's sense) the clarification and contextualisation of competing frames of value, what kind of meta-frame could best serve to facilitate integrating our knowledge about evaluation? *The answer to this question involves the recognition of the 'layered' or 'nested' nature of social reality which moves from surface levels (or explicate) phenomenon to ever deeper (implicate) more complex orders which are less amenable to the kinds of empirical inquiry which characterise 'mainstream' evaluation.*

Thus, I would at first suggest that critical realism offers a powerful way forward in the search for more *context* sensitive evaluation strategies. Its emancipatory orientation and its emphasis on evaluation as an ongoing process of testing and learning represents a convincing advance on experimentalist methodology and pragmatism. However, it ignores or dismisses two important modes of evaluation which are quite compatible with the critical realism frame: hermeneutic evaluation through the price system and through the political system. Hayek has much in common with the critical realist turn in economics (see Lawson, 1997, for example) and an evaluation strategy framed by critical realism could allow for the fact that markets in a Hayekian sense can be viewed as real world experiments in what works in specific contexts. At the same time, critical realism is also rather too dismissive of the 'constructivist' approach to evaluation. The deeper we delve into the social order the more do we encounter the implicate rules and structures, values, norms, and meanings that get unfolded and made explicate in human practice, behaviour and problems. The interpretive frame with its emphasis on learning, and empowerment and understanding language and the production of meaning is, I would contend, far more in keeping with the critical realist project than Pawson and Tilley would allow. If we accept the critical realist position that there are 'levels' of social reality, then we might deploy mainstream evaluative strategies for the 'surface' level of complex problems whilst at the same time recognising their profound limitations when it comes to understanding the particular institutional and assumptive contexts within which complex problems arise and are embedded.

Fischer's *Evaluating Public Policy* is an attempt to move beyond a critique of mainstream policy analysis, which also takes a 'layered' view of the problem of evaluation, though from an interpretivist, rather than critical realist perspective. Fischer sets out to show how empirical and normative approaches to evaluation can inform one another. As he argues:

> Whereas the traditional goal of the social sciences has been to empirically discover and predict recurring regularities in the structure and processes of political behavior, the purpose of a transformational social science is to assist political actors in their own efforts to discursively understand the ways in which they make and remake their social and political systems. Such a social science must clarify and theorize about processes – both intellectual and material – through which political actors form, function within, dissolve, and restructure political worlds. To be sure, empirical research is important to such an inquiry. But its importance lies in its ability to inform a larger normative deliberation, not in its empirical predictive powers per se (Fischer, 1995, xi).

Fischer aims to show how empirical and normative policy evaluation can be integrated through a multi-methodological strategy. He suggests that evaluation can be understood on two levels: first and second order.

First-Order Evaluation

First-order evaluation includes a technical-analytical discourse that relates to programme verification and a contextual discourse that relates to situational validation.

Programme verification is evaluation concerned with questions relating to how a given policy has fulfilled its stated objectives. Here evaluation is about measurement, causality and outcome. Under situational validation, on the other hand, we are interested in asking questions about how objectives are relevant to a specific problem situation.

At the first level we are in the domain of stated defined objectives and rationale. The analysis proceeds by investigating how given goals have been met. This is the domain of 'rational', 'positivistic' or 'causal' verification: it is the world of the policy cycle, where decision making can be broken up into neat stages. However, as we begin to move towards the second level, analysis becomes more focused on qualitative issues than empirical or quantitative research. We move from 'explanation' to 'understanding', that is to say from a preoccupation with 'causal adequacy' to 'adequacy of meaning'. At the level of contextual discourse we are shifting into a more interpretive mode: we are, in short, more involved in understanding a given policy or programme program in its specific contexts. As Fischer explains:

> Whereas the empirical approach conceptualizes individuals as abstract behavioral objects for the testing of hypotheses about causal linkages, interpretive approaches treat individuals as subjective agents acting in social situations in pursuit of their goals and objectives (Fischer, 1995, p.71).

Thus at this level, as Fischer points out, analysis is grounded in theories of social phenomenology, symbolic interactionism and ethnomethodology. Evaluation in terms of situational validation is therefore to do with the 'logic' of given situation and is designed to:

> get inside the actor's situation to understand the actor's own subjective interpretations of the situation. Interpretive social science is concerned with the social actor's cognitive perception of reality: its task is to explicate empirically the actor's experientially based subjective framework which determines his or her problem definitions and the social actions based on them (Fischer, 1995, p.77).

Here the method to be used is qualitative and 'naturalistic' (Guba and Lincoln, 1990), rather than quantitative and positivistic.

Second-Order Evaluation

Second-order evaluation includes a systems discourse or societal vindication and an ideological discourse or social choice.

We enter into the second order recognising that verification and validation may well fail to settle or resolve arguments about policy evaluation. At the level of 'societal vindication' we turn from the concrete situational context to the societal system as a whole. First order analysis is concerned with those empirical and normative judgements made in given contexts about how programme objectives and goals are implemented. Second-order societal vindication, however, takes a much wider perspective in asking questions to do with the normative implications and consequences for the social system as whole. That is to say, we focus on the impact of a given programme at the 'macro' rather than the 'micro' level.

Societal vindication, therefore, is a level of analysis which is focused on the intertwining (or enfolding) of programmes into the system as a whole. Here we are asking questions about the 'fundamental' ideas, values and beliefs which operate in a society and which organise the social order. The aim is to 'tease out the value implications of policy arguments' so as to show how ideological discourse 'structures and restructures the world in which we live' (Fischer, 1995, p.22). In examining the value choices in second order evaluation we are concerned with coming to grips with the 'deeper level of assumptions which underlies and shapes our conception and identification of policy problems' (Fischer, 1995, p.171) At the level of second-order evaluation we move from seeking to *explain* causes, to *understanding meanings*. This involves a shift from the methodology of the first-order to a more hermeneutic-interpretive approach, and a move from instrumental rationality to Habermasian communicative rationality. Here evaluation is about gaining acceptability and wider social validation, rather than measurement, and calculation of costs and benefits, impacts, risk, etc.

Towards a Possible Integration

The strength of the Fischer model is that it provides a convincing framework for integrating both the empirical and normative approaches to policy evaluation. It shows how the concept of an explicate (empirical) order can be understood in relation to a more implicate (normative) order, and how more *holistic* evaluative strategies could proceed in order to integrate policy relevant knowledge. In order to do this Fischer takes several case studies: Project Head Start, the Times Square re-development plan, disability policy, and environmental risk assessment. The implications he draws from applying his framework to these case studies is that policy deliberation methodology holds out the possibility of transforming political structures and democratising evaluation which has been for so long dominated by the positivistic and technocratic frames.

> Traditional policy analysis, as we have seen, emerged as a discipline to guide the decision making processes of a society governed by large-scale techno-managerial institutions. In sharp contrast, a vigorous democracy ultimately requires a more participatory set of institutions and methods (...) A

deliberative model of evaluation would broaden the goal to include an assessment of the political needs and interests of the larger political community, rather than emphasizing the technical efficiency of the governing institutions as an end unto itself (Fischer, 1995, p.217).

Policy evaluation in a deliberative sense becomes focused on ideas and arguments and is essentially a communicative enterprise. The expert is placed in a participative relationship with clients and citizens.

A participatory policy analysis (...) would gear expert practices to the requirements of political empowerment. Rather than providing technical answers designed to bring political discussion to an end, the task would be to assist citizens in their efforts to examine their *own* interests and to make their own decisions. The expert would be defined as a 'facilitator' of public learning and empowerment. Beyond merely providing data, the facilitator would seek to integrate process evaluation with the empirical requirements of technical evaluation. As such the facilitator would become an expert in how people learn, clarify and decide for themselves (Fischer, 1995, p.222).

Fischer's work is really seeking to explore what kind of institutional re-design needs to take place in order to improve democracy in the knowledge society. Evaluation grounded in a more hermeneutic or *interpretive* tradition, he believes, is the approach which can serve to both extend our understanding of the policy process and policy problems and enlarge the possibilities of democratic governance. This idea of evaluation taking place in the context of *orders* or levels offers a way forward to understanding how the design of evaluation strategies should aim to combine the empirical and the more qualitative and interpretivist approaches. Understanding causes, and exploring meanings are not necessarily incommensurate when we realise that different *orders* of the social world require different kinds of approach. This idea of evaluation taking on board the stratification of the social world echoes the kind of arguments used by critical realist approaches. If we place the Fischer model in the context of the ontological position of the critical realist approach (Figure 5.3) we can argue that frames E1,2,3,4, and 5 constitute 'first order' empirical approaches which are concerned with the 'empirical world' or 'domain' (from the point of view of E8).

Let us extend Fischer's sense of second order evaluative frames to include E7, that is evaluation through markets and prices alongside interpretive evaluation. The focus of hermeneutic evaluation is really on meaning, values, tacit knowledge, language, policy or programme theories.

It is concerned with what we might term the assumptive world. From the point of view of critical realism, the 'empirical world' is the outcome or unfolding of the 'real domain'. Which, from a realist reading of Hayek, could be said to be the focus of his approach. In turn, the process of testing CMO configurations – context, mechanism and outcome – could be said to contribute to the learning process and the enhancement of human freedom.

The key dimension of Fischer's model is the process whereby first order claims to knowledge are subjected to deliberative and participative testing. In the realm of second order evaluation knowledge, assumptions and values are explored and challenged through a participatory evaluation strategy. *This does not mean we abandon the 'empirically' based methods of mainstream evaluation, but rather create processes, and institutions whereby competing and oftentimes complementary evaluation frames can be the object of deliberation and learning* (see Parsons, 2000). As in the case of the Dutch transport plan, for example, different ways of valuing transport problems and solutions can be made more tractable by making their assumptions clearer.

Figure 5.3   First and Second Order Evaluative Frames

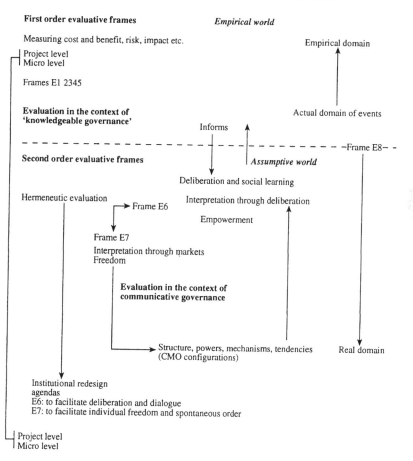

There is, in other words, no need to throw out babies E 1,2,3,4 and 5 with the positivist bath water. The process of *valuing* requires appropriate democratic *venues*. For the disciples of Hayek, of course, that venue is ultimately the market place. Interpretivistic strategies, on the other hand, do not mean that we should abandon the quest for more 'evidence based policy research' – quite the opposite. But, from Fischer's point of view, evaluation needs to recognise that 'facts' cannot resolve matters: evaluation is a political process and requires the redesign of the policy process as a whole so as to take account of – not ignore – contested meaning. The need for new evaluative processes is also the message of the critical realist frame: evaluation ought to be so re-designed so as to better facilitate the testing of CMO configurations and promote learning institutions and a learning society. And, from the point of view of the Hayekian approach, local evaluations, derived from local knowledge (of costs and benefits) are necessary: but these local evaluations cannot and should not be aggregated. It is for markets and the voluntary planning process to learn from evaluation.

By situating mainstream evaluation methods in transport in the context of critical frames, therefore, evaluators (and their clients and citizens) could better understand the strengths and weaknesses of analytical tools and policy instruments. Indeed, it may be that the concept of evaluation as a learning process could provide a fitting context within which to reflect on the integration of evaluation frames, in transport as in other policy areas. This reflectiveness should extend into developing methodologies for actually integrating the *results* of multi-framed evaluation projects. The aim would be to ensure that policy makers and evaluators are able to understand where they are coming from, and to be prepared to 'put themselves in the shoes' (Schön and Rein, 1994, p.187) of those coming from different assumptive worlds to their own.

Putting Theory into Practice

A methodology which has, perhaps, a great potential for advancing the kind of strategy for transport evaluation suggested by Fischer is the idea of making use of the 'art of strategic conversation' or scenario planning. Scenario planning is an approach developed by Royal Dutch Shell in the 1970s and 1980s and has been widely used. As the GBN (Global Business Network) describes it:

Scenarios are tools for ordering one's perceptions about alternative future environments in which today's decisions may be played out. In practice scenarios resemble a set of stories (...) built around carefully constructed plots. Stories are an old way of organizing knowledge; when used as strategic tools they confront denial by encouraging – in fact, requiring multiple perspectives on complex events (...) Stories are powerful planning tools precisely because the future is unpredictable. Unlike traditional forecasting or market research, scenarios present alternative images instead of extrapolating trends from the present. Scenarios also embrace qualitative perspectives and the potential for sharp discontinuities that econometric models exclude (...) scenarios provide a common vocabulary and an effective basis for communicating complex – sometimes paradoxical – conditions and options (...) Ultimately, the result of scenario planning is not a more accurate picture of the tomorrow but better thinking and an ongoing strategic conversation about the future (GBN, 2000).

A key aspect of scenario planning is to place decision making in conditions of risk and to suggest ways in which we can deal with uncertainty. In transport as in other areas of policy making, the internal and external environment are subject to considerable unpredictability and indeterminism. As Heijden explains:

(...) forecasting and scenario planning have very different purposes. The strategic question has its origin in uncertainty, both in the environment and within the organization. Uncertainty increases the further we look. Forecasting is useful in the short term, where things are reasonably predictable and uncertainty is relatively small compared to our ability to predict. In this range rationalistic 'predict and control' planning makes sense and is necessary. In the very long terms where very little is predictable planning is not a useful activity. It is in the intermediate future where uncertainty and predictability are both significant that scenario planning makes its contribution (Heijden, 1996, p.106).

How organisations respond to uncertainty will be shaped by the mental maps which exist to frame and make sense of reality. A starting point for the methodology is what individuals and groups perceive is shaped by the schemas and concepts they use. It follows that if we want to think more creatively we have to find ways of overcoming the impetus towards one track, linear ways of looking at the world.

Scenario planning proceeds by recognising the primacy of uncertainty, and acknowledging the difficulties of measuring and control. The key issue for scenario planning is how we can become clear about our options and values in various (plausible) stories of the future as an envelope of possibilities. The approach stresses the need for thinking (strategically) about 'futures' rather than a single 'forecast' of where the future is going to be. Scenario planning is consequently an art which rejects the notion of linear, straight line forecasting and builds on the idea that we cannot know the future, so much as we can tell stories about the future. Whereas linear forecasting is about projecting the past into the future and extrapolating from what we know in the present into what we see as being the future, scenario planning is predicated on a *multiplist* view of the relationship between human beings and the future. For scenario planning the future does not exist 'out there' but is contained in the stories we tell about it. The future is to be understood less in *deterministic* than in *deliberative* terms. The history of traffic forecasting shows well enough that we cannot predict the future: but we can better understand the stories and myths we tell about it and the values which shape the future.

Heijden argues that one of the main benefits of the use of scenario stories is that they provide a way of organising complex and 'seemingly unrelated data' (Heijden, 1996, p.106) so that events and problems can be understood in a broader context. In other words, stories become a language within the organisation: a way of thinking about the future and what action is necessary to navigate with success in an uncertain world. Above all, scenarios should be viewed as ways in which people and organisations can learn. As the world is uncertain and as we can predict so very little, it makes sense to say that the art of planning in such conditions is to devise ways in which we can learn to navigate without clear, well drawn precise maps. Scenario planning is planning for a fuzzy world in which learning to learn (planning as learning/evaluation as learning) is the key to success. The idea of constructing scenarios, therefore, is very different to the notion of trying to 'forecast' the future in a probabilistic way. It lays no claim to being objective in a simplistic sense but recognises that our knowledge of the future is essentially of a participatory kind. Arie de Geus makes the point that scenarios are not a superior form of planing, so much as:

> Tools for foresight: discussions and documents whose purpose is not a prediction or a plan, but a change in the mindset of the people who use them (...) [and] Ironically enough, a de-emphasis on prediction seems to lead to accuracy about the future (de Geus, 1999, pp.59-67).

In integrating evaluative frames, therefore, we should keep in mind what we can learn from the work done at Shell and other organisations to promote forms of social and economic planning which can facilitate a sense of participation in the way in which we plan, forecast, set targets and evaluate. Bonnett and Olson (1998), for example, demonstrate how scenario planning greatly enriched public policy decision making in the case of the development of New Jersey's Long-range State-wide Transportation Plan. It served, they argue, to help participants:

> challenge conventional wisdom about recent trends, articulate their own expectations, and be critical of conventional assumptions about the future. The process of building scenarios is an important learning opportunity for organizational leaders. Direct involvement in scenario development forces leaders to think critically about plausible futures and appropriate organizational strategies. It forces leaders to understand and articulate some of the implicit assumptions they have about what the future will be like and why they think that (Bonnet and Olson, 1998, p.309).

Citizen groups in New Jersey were a core and integral part of the development of the various scenarios and the evaluation of policy choices. A scenario planning approach to transport evaluation would be about seeking to map out the various *stories* which competing frames would tell about a given problem. It may be, for example (indeed it *will* be the case) that no one story can fully capture what the impacts will be (*ex ante*) or what the impacts have been (*ex post*) of a given policy, programme or project. Different stories about transport futures will elicit different policy, programme and project possibilities. In scenario planning the aim is to bring these differing stories into an order which allows a conversation to take place.

Above all, the non-mainstream frames raise the profile of the vitally important issue of how transport evaluation has to be more engaged with the issues of public acceptability and social and political validation. If these arguments are taken on board by mainstream approaches this would involve evaluative strategies in transport using techniques like scenario planning, allied to citizens forums, user panels, citizen's juries, planning cells etc.,[5] which could improve the sensitivity of (mainstream) evaluation to the need for greater social and political acceptability amongst stakeholders and citizens. Evaluative designs, therefore, should seek to challenge and illuminate the (oftentimes deep-seated) attitudes and assumptions underpinning the transport stories which people tell and the mental maps which come to frame the decision making processes. The message of Fischer and Schön and Rein is that we have to find ways of contextualising the range of frames (stories) which may be deployed in the policy process in general, and in evaluation in particular. This might seem rather idealistic and impractical but the evidence from the use of scenario planning by some of the world's most successful organisations suggests that interpretive (story telling) approaches can be utilised for the kind of strategic decision making which characterises transport issues. And, keeping in mind the kind of arguments deployed by the ETMS frame, even though we cannot know or predict, scenario methodology suggests a way of making decisions in conditions of complexity and uncertainty that does not proceed with the MOE assumptions of mainstream evaluative frames, and yet arrives at a way of planning (*qua* learning) which is (I would argue) in line with some of the central MOE assumptions of the Hayekian frame.[6]

A scenario approach to the problem of transport evaluation might begin, therefore, by recognising that there are a multiplicity of evaluative stories (or what I refer to as frames). The task of integration would, in practical terms, involve taking specific policies, programmes and projects and allow the competing frames to tell their respective stories as to how they see what has happened, or what might happen. Then, these stories should serve as the basis for 'strategic conversations' to determine which stories, forecasts, visions are most plausible and acceptable. And, as in the case of the New Jersey exercise, stakeholders and citizen's panels should be involved in the process of envisioning possible transport scenarios and possible responses.

Perhaps the development of more pluralistic and democratic evaluative strategies are not such fanciful propositions. Terry, for example, makes the point when arguing the case for more evidenced based transport policy that, in Britain:

> For a long period, evidence from within, or commissioned by the DoT (Department of Transport) was the only evidence admissible in policy making. It was not until the 1990s, when organisations outside government (...) successfully challenged the traditional direction of transport policy, that alternative evidence was given much weight. (...) The dissemination of evidence about 'what works' in transport policy has arguably become more pluralistic in the 1990s as compared to the 1970s and 1980s (Terry, 2000, p.202).

This implies that transport evaluation (as other policy areas) may well become more open to alternative ways of thinking, valuing and interpreting goals and ways of achieving them. In turn this means focusing on the central issue of how knowledge, information and evidence is utilised in the policy process and how different frames can be usefully integrated.

Of course, such a (paradigm) shift in the design of transport evaluation – a field which has been so very dominated by the mainstream frames – would be inherently subversive as it would bring to light the tacit or unconscious values, theories, perceptions, assumptions and beliefs in policy and research designs which it might be political to keep vague and ill-defined. But, of course, that was precisely *why* Lasswell believed that a policy science which integrated knowledge through the clarification of values was such a vital dimension of extending and improving democracy. In a democracy we should take nothing for granted: least of all the assumptions of policy makers and those whose business it is to process and produce knowledge about a public problem as complex as that of transport.

Notes

1   Specifically so called 'Q' methodology.
2   This section is based on Michel van Eeten's report, published in Dutch in 1999 (Van Eeten, 1999a). Not being a reader of Dutch I am relying on an account of the project as presented in an unpublished paper by Eeten (Van Eeten, 1999b).
3   Roy Bhaskar, one of the leading exponents of critical realism, has sought to explore some of the central problems of positivist ideas of science and to advance an account of science in terms of 'transcendental realism'. His focus, however, is very different to the majority of philosophers of science in that he is concerned with ontology: specifically how epistemologies of science involve implicit ontological assumptions. The conventional, or mainstream idea of science employs an ontology which may be termed empirical realism. 'Underpinning empirical realism is a model of man in which men are seen as sensors of given facts and recorders of their constant conjunctions: passive spectators of a given world rather than as active agents in a complex one' (Bhaskar, 1997, p.198). This presents science with a problem: an epistemic fallacy, since what is real (event regularities) and what can be sensed, observed and 'known', is confused with what is real. Questions of ontology are, he maintains, reduced to those of epistemology. In turn this gives rise to what he terms an ontological tension (Bhaskar, 1989, p.18) as between the way science may be characterised – as to do with event regularities – with the practice of science which is about seeking to understand the causal mechanisms which generate the empirical phenomena. Furthermore, in seeking to understand the causal mechanisms, scientific experimentation involves isolating the effects of a given causal mechanism so as to close off the effect from all other causal powers, structures and mechanisms. This 'closure' is, he contends, crucial to the claims for a realist account of science. Thus science may be said to be about 'real' 'facts' but these real facts are essentially a social product and 'historically specific social realities'. How, then, can we understand science when a given phenomenon is placed back in the open, non-experimental environment? Bhaskar's response to this

problem is 'transcendental realism'. A realist account of science has to address the ontological distinction between the observable patterns and event regularities, on the one hand and the causal laws (generative structures, capacities, causal powers, mechanisms) on the other. The latter, causal mechanisms, are, for Bhaskar, the intransitive objects of scientific investigation: these may not be empirically observable. The event regularities or empirical phenomena are transitive: it is these transitive phenomena which are the object of (closed) experimentation. Outside the rare and special conditions of the closed environment of the experiment, however, there are numerous causal forces, or 'tendencies' that interact in complex ways. The story of science is the story of how, through the transitive and socially contingent domain of experimental derived knowledge, we have gained understanding of the domain of the intransitive which, unlike our investigations themselves, is independent of us. Transcendental realism accounts for science as an activity which is both the product of society, history and our theories, and a quest to understand non-observable causal mechanisms. The world, therefore, may be viewed as layered into three domains: empirical, actual and real. What we observe as empirical phenomena are the outcome of highly complex and contingent causal mechanisms which are not reducible to their various components, and may not be directly accessible to us. The 'real' world gives rise to actual events and states, but the real may not always become actual, and the actual may not be observable and empirical. Furthermore experiences, events and mechanisms are out of phase with one another, thus giving rise to different experiences of the same events. As in the case of watching a game of football, events are experienced differently: 'Events, in other words, are conjointly determined by various, perhaps countervailing, influences so that governing causes, though necessarily 'appearing' through, or in events can rarely be read off' (Lawson, 1997, p.22). Science, from the transcendentalist point of view, therefore, is perceived as essentially a social and human product which is defined less by the regularity of states of affairs than a mode of socially produced knowledge concerned with the mechanisms which generate or give rise to the empirical world. For the critical realists, therefore, social science should be focused on understanding and explaining the deep structures of social life: structures which, unlike those of the natural world, both facilitate and depend on human action. Empirical approaches which are concerned with the explanation of social, political, economic and other forms of human life, in terms of transitive, surface event regularities are thus viewed as inadequate and unscientific.

4   For a discussion of Hirchman's ideas on exit voice and loyalty, see Parsons, 1995, pp. 525-527.

5   On the issue of widening citizen participation in public policy, see Khan (ed.), 1999.

6   Significantly, the kind of ontology/epistemology used by scenario planners is 'layered' rather like the critical realist model. Heijden, for example, argues that: 'Scenario planners train themselves to find structure in a range of events. One useful way of thinking about the data is through the categorisation known as the 'iceberg', which breaks down knowledge into three categories: events, pattern and structure. At the top of the iceberg, 'above the surface' are visible events. For example developments in the market or customers doing one thing or another, governments enacting legislation, and so on. Events can be observed. One sees the world though events that present themselves and that we perceive. It is the visible part of the iceberg. Initially we describe the world in those terms. But as soon as important

events present themselves we try to discover underlying patterns and structure, in order to understand the situation (...) Scenario planners start from the premise that there is much more to be said than just reporting events. They assume structure underneath events' (Heijden, 1996, pp.97-8).

References

Anderson , T.L. and Leal, D.R. (1991), *Free Market Environmentalism*, Pacific Institute for Public Policy, Westview Press, Boulder, San Francisco.

Bhaskar, R. (1989), *The Possibilities of Naturalism: A Philosophical Critique of Contemporary Human Sciences*, Harvester, New York.

Bhaskar, R. (1997), *A Realist Theory of Science*, Verso, London.

Bonnet, T.W. and Olson, R.L. (1998), 'How Scenarios Enrich Public Policy Decisions', in Fahey, L. and Randall, R.M. (eds), *Learning From the Future: Competitive foresight Scenarios*, John Willey, New York.

Brown, J.S. and Duguid, P. (2000), *The Social Life of Information*, Harvard Business School Press, Boston.

Burczak, T. (1994), 'The Post-Modern Moments in F.A. Hayek's Economics', *Economics and Philosophy*, Vol. 10, pp.31-58.

De Geus, A. (1999), *The Living Company*, Nicholas Brealey Publishing, London.

Dryzek, J.S. (1997), *The Politics of the Earth: Environmental Discourses*, OUP, Oxford.

Fischer, F. (1995), *Evaluating Public Policy*, Nelson Hall, Chicago.

GBN, Global Business Network (2000) 'Scenarios',
    http: // www.gbn.org/ public/gbnstory/scenarios

Guba, E.G. and Lincoln, Y. (eds) (1990), *The Paradigm Dialogue*, Sage, London.

Harvey, D.L. and Reed, M. (1996), 'Social Science as the Study of Complex Systems', in Kiel, L.D. and Elliott, E. (eds), *Chaos Theory in the Social Sciences*, University of Michigan Press, Ann Arbor, pp. 295-323.

Hayek, F.A. (1944), *The Road to Serfdom*, Routledge, London.

Hayek, F.A. (1948), *Individualism and the Economic Order*, Chicago University Press, Chicago.

Hayek, F.A. (1948a), 'The Meaning of Competition', in Hayek, F.A., *Individualism and the Economic Order*, Chicago University Press, Chicago.

Hayek, F.A. (1952), *The Sensory Order*, Chicago University Press, Chicago.

Hayek, F.A. (1957), *The Counter-Revolution in Science*, Routledge, London.

Hayek, F.A. (1960), *The Constitution of Liberty*, Routledge, London.

Heijden, K. Van der (1996), *Scenarios: The Art of Strategic Conversation*, Wiley, New York.

Hirchman, A. (1970), *Exit, Voice and Loyalty*, Harvard University Press, Cambridge.

Kahn, U. (1999), *Participation Beyond the Ballot Box: European Case Studies in State-Citizen Political Dialogue*, UCL Press, London.

Kiel, L.D. and Elliott, E. (eds) (1997), *Chaos Theory in the Social Sciences*, University of Michigan Press, Ann Arbor.

Klein, D.B. (1998), 'Planning and the two Coordinations', *Planning and Markets*, Vol. 1, No. 1, pp. 1-9.

Lasswell, H.D. (1958), *Politics: Who Gets What When and How*, Meriden Book, Cleveland, Ohio.

Lasswell, H.D. (1970), 'The Emerging Conception of the Policy Sciences', *Policy Sciences*, Vol. 1, No. 3, pp. 3-14.

Lawson, T. (1997), *Economics and Reality*, Routledge, London.

Parsons, W. (1995), *Public Policy*, Cheltenham, Edward Elgar.

Parsons, W. (2000), *Public Policy as Public Learning*, Queen Mary, University of London, London.

Pawson, R. and Tilley, N. (1997), *Realistic Evaluation*, London, Sage.

Pennington, M. (2001), 'Collaborative Planning: a Liberal Critique', in Allendinger, P. and M. Tewdr-Jones, *Planning Futures*, Athlone Press, London (forthcoming).

Rittel, H.W.J and Webber, M.M. (1973), 'Dilemmas in a General Theory of Planning', *Policy Sciences*, Vol. 4, No. 2, pp. 155-169.

Schön, D.A. and Rein, M. (1994), *Frame Reflection: Toward the Resolution of Intractable Policy Controversies*, Basic Books, New York.

Terry, F. (2000), 'Transport: Beyond Predict and Provide', in Davies, H.T.O., Nutley, S.M. and Smith, P.C., *What Works? Evidence-Based Policy and Practice in Public Services*, The Policy Press, Bristol, pp.187-205.

Van Eeten, M. (1999a), *Perspectieven op verkeer en vervoer en hun consequenties voor agendavorming*, rapport aan het Ministerie van Verkeer en Waterstaadt, Delft, Technische Bestuurkunde.

Van Eeten, M. (1999b), 'Recasting the Hard Issues through Discourse Analysis: The Case of Transportation Policy in the Netherlands', Unpublished paper for the *Symposium on Theory, Policy and Society*, University of Leiden, June, 24-25, 1999.

Young, K. (1977), 'Values in the Policy Process', *Policy and Politics*, Vol. 5, pp. 1-22.

6 Evaluation of Projects and Programmes: Principles and Examples

FRANK HAIGHT

Project evaluation, in my view, is balanced somewhere between a science and an art, that is, between fact and opinion. In the best of circumstances, each should re-enforce the other: opinion should be based on fact of course, but also the facts are usually based on someone's opinion – opinion of what is important.

It is remarkable that, despite differences in date, in language and in provenance, evaluation handbooks cover so much of exactly the same ground. Let us begin with some simple vocabulary for a simple evaluation. There is a project that we are called upon to evaluate. We will be reporting to an administrator of the project. The project has experienced an intervention that the administrator wishes us to evaluate, and has designated some parameters of interest. There is a goal that we are striving for: has the intervention been successful, as shown by the parameters, or, more frequently, how successful has the intervention been, again based on the performance of the parameters? In other words, 'did it work?' or 'how well did it work?'. You will notice that I have used the singular ('goal'); this is what I call a simple evaluation.

Unfortunately, none of these expressions has achieved universal acceptance. Often, for example, the project is called the programme. Some others and I will reserve that term for a series or collection of projects. The parameters are, or were, called the MOE, or measure of effectiveness or outcome measures by some agencies. The administrator is often called the agency; in any case it is the boss, to whom we evaluators must report. Thus, the administrator commissions the evaluator to examine the agency objective, and the evaluator has a goal, usually to provide the administrator with accurate and reliable information about the project.

This paper deals mostly with evaluation in general, although many of the examples I will use are drawn from the field of traffic safety. These

181

examples, in turn, reflect to some extent the classification and vocabulary used by the United State Department of Transportation, more specifically by the Federal Highway Administration (FHWA) and the National Highway Traffic Safety Administration. The principles involved, on the other hand, are intended to apply to many – if not most – situations where evaluation is being proposed.

There are two important limitations to my discussion. First, many projects in traffic safety and I suppose in other fields as well, have a political component, which is so important as to make the scientific methods nearly irrelevant. Secondly a special type of political constraint relates to money; I must avoid dealing too dogmatically with funding. It is not really fair to expect academic analysis to say where or in what quantity public – or for that matter private – money should be spent. The best we can do is to offer principles and alternatives.

The protection of the environment provides a good example of both limitations. Until we are told which aspect of the environment, what options can be considered to protect it, which agencies should be responsible, what constraints – especially financial constraints – are necessary, and to do this in numerical terms, it will be quite difficult for any scientific analysis of evaluation methods to reach firm conclusions.

Project evaluation, in the sense of controlled experimentation and analysis, is relatively new. Some engineering projects, especially in the pre-1960s period, were considered to contribute to safety on the road without careful statistical studies of effectiveness: stop signs, traffic lights, divided highways, etc. were considered so necessary for the development of transport as to be self justifying. The characteristic view (and one widely accepted) was that a safety project was justified 'if it saved even one life'.

By the mid-1960s, however, that particular slogan was generally recognised to contain at least two fallacies: first, there was no reliable way to tell if a project did in fact save one life; second, even if a project was demonstrably cost-effective, there remained no guarantee that other programmes might not be equally effective at lower cost, or more effective at equal cost. In addition, by the simple act of adjusting the value of what was called a 'statistical life' virtually any project could be found to be cost effective. Such a simple example illustrates the need to specify quite exactly the parameters of the problem if a satisfactory evaluation is to be performed.

These parameters are in turn often subject to political pressures of one sort or another. If an evaluation is implicitly designed to satisfy some such pressure, including pressure to justify decisions already made, it could be called a 'pseudo-evaluation' and will not be discussed further. It should be admitted that some handbooks justify pseudo-evaluation. Decision-makers may wish to select a project for evaluation to confirm their impression that it is performing well. If the evaluation confirms their opinions, elected officials or top management may be able to satisfy critics of an effective although unpopular project.

A temptation that must be resisted is to specify remedial interventions. It is one thing to evaluate whether some parameter has been reduced say from four to three by the given project. It would be unwise to proceed to tell how to reduce it still further; that should remain a political decision.

It is particularly true in developing countries that overseas consultants tend to assume the effectiveness of countermeasures, which have been successful in their home countries. As an example, a Finnish team which produced a thorough and admirable study of road safety in Kenya (Viatek, 1980), observing that '[T]here are many perfectly dangerous places in the large cities of Kenya, where dense pedestrian flows are crossing streets with busy vehicular traffic', states that '[A]n effective remedial countermeasure is the construction of under- or overpasses'.

Note that there is no requirement that Kenyans be urged or even compelled to use the overpasses, only that they be constructed. The assumption seems to be that Kenyans, like Finns, would make use of the overpasses, and that the cities of Kenya would necessarily become safer.

When to Evaluate

In planning a project, it is important to make an early decision as to whether or not the project deserves to be evaluated. Just as there are certain projects that should be evaluated, so there are others that should not.

Evaluation is properly regarded as an aid to decision-making, and is normally performed only in circumstances where some future decisions are envisaged. Being a difficult and often costly task, evaluation should not be undertaken for purely frivolous reasons, such as justification for decisions already made.

There are a number of factors (or 'principles') which should be taken into account in making the choice whether or not to evaluate a project. A first consideration is the nature and importance of the future decisions. If the project is a pilot study, it is likely that the outcome will influence many similar projects in the future, and so would be a strong candidate for evaluation. On the other hand, a 'one shot' project would be less likely to deserve evaluation.

A particular case where evaluation would have a persuasive rationale occurs when alternatives are being considered: different road geometries, different public information plans, different land use arrangements. Many interventions can be implemented in a variety of strengths and configurations, and in many cases, it is worthwhile running carefully designed (and evaluated) pilot projects.

A second factor to consider is the cost of evaluation. The U.S. National Highway Traffic Safety Administration manual offers a rule of thumb that 'evaluation costs should not exceed 10 per cent to 25 per cent of the project cost unless the project is very special'. In preparing the budget for the project, the evaluation costs should be considered as carefully as the costs of the project itself.

If adequate money and manpower cannot be allocated to evaluation, it may be wiser to let the project proceed without evaluation. After all, a well-designed project may have a persuasive rationale in and of itself. It has been observed that the advice given school children to 'look in both directions before crossing the road' has not been carefully evaluated, and probably does not need to be. It would be a mistake to assume that a project can be successfully evaluated on a shoestring budget. There are some cases, such as pilot projects which, if successful, will be widely copied, where a first class evaluation can be expected to cost as much as or more than the implementation.

A third consideration must be the chance of a reliable conclusion. If an evaluation yields data which is not statistically significant, or which is unreliable in some other way, or which cannot be reliably tied to the intervention in question, the exercise may be a waste of money.

This risk is greatest in dealing with imponderable effects, such as those emanating from a public relations campaign. Social action projects often have vague or ill-defined objectives, nearly impossible to evaluate objectively. It is less so in interventions, which are specific, such as the reconstruction of a particular intersection. But even in these cases, it is necessary to assess carefully the chance of a conclusive outcome. Will the necessary data be available? At what cost? Will the data be reliable? What will be the cost of coding and storing the data? What are the chances that the sample size will be so small as to give an ambiguous answer? How long will it take to accumulate an adequate sample? What other factors might be influencing the evaluation during the experimental period? What specific problems might be encountered in forming a conclusion? What level of significance will be needed for the conclusion to be useful? These three issues: the need to know, the cost of knowing, and the chances of success, are combined to produce a decision regarding the desirability of proceeding with an evaluation.

Let me give you an example of difficulties once experienced in conducting an evaluation on behalf of the U. S. Government. The topic was a reduced speed limit, to 55 m.p.h. that followed the energy panic of 1973-1974. Although this measure was designed to save fuel, there was an unexpected coincidence with a precipitous drop in the vehicle-mile fatality rate. The Nixon Administration, constrained by constitutional issues of state/federal authority and especially by the felt need to adopt measures that would be popular, relied mostly on public relations and in particular on a highly publicised 55 mph speed limit. The legacy of that approach was nearly a quarter century of wrangling involving issues having nothing to do with the energy panic, which vanished almost as soon as it came. From the point of view of evaluation, there were many complications. First, there were fifty jurisdictions, each having very slightly different laws. Second, the enforcement of these laws was partly in the hands of local government, with thousands of separate police agencies. Third, the consequence of non-compliance varied greatly, from trivial fines in some Western States, to really draconian penalties, mostly in the East. Fourth, the only really good measure of safety effectiveness was fatalities, even though, if we believe economists, fatalities are only a part of the cost of accidents. Fifth, it was not clear whether we should use fatalities per distance travelled, or fatalities per number of cars, or fatalities per capita, or any one of many other measures.

The situation was complicated at this stage by the enormous degree of popular approval of 55, even – or perhaps especially – by those who seldom obeyed it. The country was polarised into pro- and contra-advocacy that often appeared to foreclose rational discussion. A technical issue had been converted by propaganda into a moral issue, and thus into a political issue. Public opinion favoured 'saving lives' over 'saving time' and regarded the speed law as doing exactly that, even as average travelling speeds crept up to and then beyond 65.

The attempt to measure the effect of the law was surrounded by controversy, so that it was difficult to form a good experimental design, to collect data from myriad agencies, much less to reach a consensus. In the end, we could say only that 'most people' thought the law was responsible for 20 per cent up to 100 per cent of the increased safety.

Some Principles of Evaluation

It is generally agreed that, if possible, an evaluation should be considered to be an integral part of the project itself, and that the initial project plans should include evaluation plans.

> Once a new program is under way, it's usually too late for evaluation. Therefore considerations of program evaluation must be undertaken early in the game. Before a new program is inaugurated it is routine to do considerable advance planning regarding financing, personnel, organization, equipment, time tables, etc. It is absolutely essential that this advance planning also include procedures to evaluate the project's effectiveness.

The rule (or principle) that an evaluation plan must be specified in advance of implementation is a variation on a standard principle of statistical inference, and is based on simple logic. If an experiment is concluded, and sufficient data is available, various different conclusions can be supported data fishing.

An example: In a programme to suppress alcohol abuse amongst truck drivers, it would be possible to base conclusions on: (a) total number of drivers apprehended while drunk or (b) number of fatally injured drivers with excess alcohol in their blood or (c) number of trucks clustered around a popular bar or (d) number of accidents recorded during the hour after the bars close or any of these further restricted to certain age groups, to those with previous convictions, to those with previous accidents, etc. It is sometimes amusing, and always frustrating to see that police enforcement programmes are equally likely to be judged 'successful' if they either increase or decrease the number of arrests for the particular misdemeanour. Either the police are more vigilant or the public is better-behaved, either of which implies a successful intervention.

With sufficient 'fishing' it is often possible to find one measure, which indicates that a project is 'successful', but at the same time there may well be other measures, which indicate 'failure'. In the example given above, either might indicate success, a rather pleasant situation for the administrator. I can imagine that a similar situation could be devised for evaluations of other social programmes.

The decision as to which measure to adopt properly belongs in the planning stage, not after the project is completed. If the evaluator is free to choose the parameter to be used – and the value of life – after the project is completed, nearly any desired effect could be produced.

In an evaluation (as in statistical inference generally) the effect to be tested should be clearly specified in advance of the implementation of the project. The specification should be sufficiently complete so that evaluation after the implementation is confined mainly to data collection and analysis.

There are, of course, situations in which exploration of data will be appropriate; such is often the case in preliminary studies designed to formulate a hypothesis for subsequent testing. It is important to distinguish between investigations leading to the formulation of a hypothesis, and those designed strictly to test the hypothesis.

Statisticians are sometimes appalled by the bad design of experiments that they are called upon to analyse. Some experimenters copy a design from a textbook; some invent one for themselves. We are forever urging them to get a good statistical consultant at the beginning. It will usually take only an hour or two, and the consultant may have useful suggestions. It is a good prophylactic against trouble and frustration further down the line.

A second principle of evaluation is that the personnel responsible for carrying out the evaluation should as much as possible be objective and unbiased with respect to possible conclusions. Bias in the evaluation team can lead to faulty conclusions in a number of ways. In some cases, there may be a poor choice of effects to be studied, specifically a choice unrelated to the project. For example, a training programme might be evaluated by simply counting the number of persons passing through the training sessions. Or the effect specified in the original plan may be ignored if it does not yield the desired result, and substituted with another 'correct' evaluation. For example, lacking evidence of success in safety improvements, project managers sometimes cite 'improved public awareness of the problem', although improved public awareness was not one of the original purposes of the intervention. Another folly is a tendency to assume that a change in attitude is a good proxy for change in behaviour. The relationship between attitude and behaviour is complex.

There is one situation in particular where objectivity can be seriously at risk: when members of the evaluation team have a stake – financial or professional – in showing the project to be successful (or unsuccessful). As

Hatry (1980) notes this is equivalent to 'Letting those whose work is being evaluated, in effect, evaluate themselves'.

Rossi (1977) points out that

> The will to believe that their programs are effective is understandably strong among the practitioners who administer them (...) Most evaluation researches which are undertaken at the behest of the administrators of the programs involved are expected to come out with results indicating that the program is effective.

It may not always be possible to find entirely objective evaluators, but this should be the goal. A sensible path towards the goal is to assign the evaluation to individuals who are independent of the project management.

A third principle to bear in mind is that a statistical tool, like any tool, needs to be used carefully and wisely if it is to be effective. Human judgement enters into the process at several stages. Although textbooks of statistics give an idea of the wealth of tools and techniques that are available, they have little to say about how choices are made among alternative assumptions, data and methods of analysis. These crucial choices are not guided by theory, nor by intuition, but by the craft knowledge of the analyst (Majone, 1980).

In this chapter I will give attention both to the formal statistical procedures and to the development of this 'craft knowledge' of evaluation.

Types of Evaluation

It is customary to distinguish three phases of any project, namely the planning, implementation and evaluation phases.

Although these categories provide a convenient vocabulary for discussing an intervention, it would be a mistake to consider them to be either mutually exclusive or sequential.

The point has already been emphasised that certain components of evaluation should be present from the beginning: evaluation planning goes along with project planning. Evaluation is also appropriate during the implementation stages, especially if implementation is carried out over a period of time; this is often called monitoring.

Administrative Evaluation refers to the process of seeing how well the project is actually implemented. For example, if the project is designed to carry out certain changes, it is obviously desirable to ensure that the work as performed does in fact make such changes.

Impact Evaluation refers to the process of inferring the effect of the implementation on previously designated parameters. The term effectiveness evaluation is also sometimes used.

There is a necessary relationship between these two types of evaluation: No effectiveness evaluation can possibly give a positive conclusion unless there is also a positive administrative evaluation. No matter how much a parameter has improved during a project, the improvement must be specifically attributable to the intervention itself and not to some parallel but exogenous factors. A necessary first step in this verification process is a positive administrative evaluation, indicating that the intervention has taken place as intended.

Finally Clinical Evaluation refers to procedures for combining objective results with other information (e.g. from other jurisdictions, similar projects, experimental studies etc.) to obtain a balanced judgement of effects. This judgement, combined with a cost assessment, will permit a best estimate of project cost-effectiveness. Just as a physician will combine outcomes of laboratory tests and other data with his knowledge, the project evaluator should combine his knowledge with statistical conclusions. It is exactly at this juncture that objectivity and impartiality are most necessary.

There may be some difference of opinion as to whether clinical evaluation has a role to play; the concept seems to vitiate the requirement of rigid statistical procedures. However, it must be conceded that this requirement is, in many practical instances not strictly feasible; 'craft knowledge' must also play a part at all stages of the evaluation process.

In addition to these distinctions, evaluations can also be divided into two classes according to whether they are qualitative ('Has there been a significant effect on the reference parameter?' – 'Does the project work?') or quantitative ('How much has the reference parameter changed its value?' – 'How well does the project work?'). This distinction corresponds to the two branches of statistical inference: hypothesis testing and estimation. We now discuss some of the more usual experimental designs.

Before and After Design

By far the most common design is the before-after comparison. Data from the system is gathered, then a change (the intervention) is made in the system, and then corresponding data is once again obtained. The two sets of data are analysed to see whether the parameters of the system are affected by the change. Although the logic seems straightforward enough, the interpretation of the results of such a comparison is often the subject of considerable debate.

The principal difficulty is that the system after an intervention is usually different from the system before the change in ways other than the implementation of the project being evaluated. It is usually difficult if often impossible to separate the various influences on the outcome.

For this reason, it is generally believed that a before and after design is the option of last resort. There are so many threats to the validity of conclusions that some experts regard before-and-after results, in the absence of trend lines or control groups, to be more likely to be misleading than informative. Council et al for example, write that '...it provides a prime example of a design which does not control for the important threats to internal validity and thus is very vulnerable to yielding the wrong answer'.

The following example illustrates also the difficulties in depending on legislation to influence behaviour. Once, long ago and far away a law was passed, and an evaluation of its effects undertaken. This law gave pedestrians the right of way over vehicles. The number of pedestrian deaths dropped from 365 in year the law was enacted to 268 six years later. This might be evaluated as a considerable achievement, but for two embarrassing facts. First, the law was not enforced, and indeed not seriously intended to be, mostly because it went against local custom and usage. It was enacted to bring the jurisdiction 'into conformance' with external standards in competition for funding.

Secondly, the year chosen – six years later – was deliberately selected to show a decline. 'Five years later' or 'seven years later' would have produced quite different conclusions, for the pedestrian fatality totals were fluctuating considerably. Clearly, some factors other than the law were responsible for the decline in pedestrian deaths.

For evaluation, the year of comparison should have been specified in advance. There are other examples of programmes which have been proclaimed, laws passed, administrative rulings promulgated, but never enforced. Examples can often be found in multi-tier government, where the central government attempts an intervention that is largely ignored by local officials.

Nevertheless, the design is popular, not only because it is intuitively appealing, but also because it appears to be easy to analyse. Also, in monitoring for subsequent time series analysis, a before/after result appears as a preliminary quick response technique. With suitable precautions, there are some situations where this design can produce a fairly valid result.

The purpose of this design, as with all others, is to infer what would have happened without the intervention. It must be obvious from earlier discussions that it is inadmissible to assume that there would have been no change.

Threats to Validity

The first difficulty is that there may be a long-term trend in the data. If no effort is made to discover (and correct for) this trend, the observed change may erroneously be attributed to the intervention under investigation. If the data needed to discover this trend were available, trend adjustment would certainly be advised before proceeding with before and after analysis.

The second difficulty is well known in both theoretical and applied statistics, and is called 'regression to the mean'. The word 'regression' in this context does not mean statistical regression in the technical sense, but only in the popular sense of 'going back towards'. In spite of the fact that this threat to validity is well known to professionals, the problem of regression to the mean remains a serious pitfall in much analysis.

First, a few examples. Consider a series, which is fluctuating in a random manner, but without any trend. One might think of the number of heads showing when 100 coins are thrown. If the experiment of throwing the coins is repeated once a minute, it is clear that the number of heads may vary, but that the average number will remain a constant value, namely '50'. Now, suppose further that a person ignorant of probability and statistics believes that he can influence the coins, and wishes them to always show the average value '50'. When the number of heads is very large, he 'hopes for' more tails on the next throw, and when the number of tails is very large he 'hopes for' more heads next time. This person will probably be successful in fulfilling his hopes, simply because an extraordinary number of heads is difficult to exceed and is more likely, therefore, to be followed by fewer heads, going back towards the mean.

A second example: pilots training to land aircraft were punished for unusually bad landings and rewarded for exceptionally good landings. It was found that the punishment 'worked'; a very bad landing was likely to be followed by a better one. But it was also found that rewards 'failed'; one usually followed an exceptionally good landing not so good. In this example, the trend may not have been exactly zero, since the pilots were, after all, learning, but regression to the mean was still present, with large departures from the trend line likely to be followed by less extreme variation (one small consequence of regression to the mean in such a situation may be that punishment would be deemed more effective than reward in improving behaviour; but only if the series of trials are truly independent).

A third and final example: persons having three or more traffic tickets during the year were given a 'remedial treatment' of lectures on traffic safety. During the year after treatment, those in the treated group had only 87 per cent of the accidents they had in the year preceding the treatment. Is the treatment effective, or is this a case of regression to the mean?

Hauer's Correction

Hauer (1980) has given a method for correcting for regression to the mean in one specific circumstance: when the method of selection for treatment consists in choosing all those sites, which have measures over a specified threshold value. There is also a technical assumption needed: that the statistical distribution of the items forming the threshold be distributed according to the Poisson probability law. If these conditions are fulfilled, then the expected number after the treatment (assuming equal time periods before and after) is not the total number before, but rather the total number that there would have been before, if the threshold had been one unit higher.

Hauer gives the example of road sections, which accumulated 15 or more accidents being selected for reconstruction. There were seven such road sections, which had 15, 15, 16, 18, 18, 23, 27 accidents respectively. The total number before reconstruction is thus 132. Using Hauer's correction, the number of accidents expected after the reconstruction is 102, i.e. the sum of accidents there would have been before, if the threshold had been 16, i.e. one unit higher than the actual threshold of 15. Hauer continues this hypothetical example by supposing that after treatment (and in an equal time period) the number of accidents was 95. This would lead to the conclusion of a seven per cent improvement due to the reconstruction (i.e. (102-95)/102) rather than the 'obvious' (but incorrect!) conclusion of 28 per cent improvement (i.e. (132-95)/132).

Sometimes the effect of the correction can be quite startling. In the example provided above with regard to driver improvement, suppose that there were one thousand drivers treated, of whom 500 had three citations, 400 had four citations and the remaining 100 had five citations. Then the expected number of citations after treatment (assuming that the treatment had no effect) would be 2,100 rather than the number before treatment, 3,600. If the result were indeed a reduction from 3,600 to 2,100, it would take a good deal of knowledge, honesty and self-control on the part of the programme administrator to keep from proclaiming the treatment programme a success. But the plain fact is the result does not give any indication of the success of the project. The reason, of course, depends on the fact that those given the treatment were not a random selection of drivers, but a very specially selected group whose record would be expected (regression to the mean) to improve without any intervention at all.

In such a situation, the better strategy would be to have divided the 1,000 offenders into a treatment group and a control group and to have compared the effects according to the methods to be discussed later.

It is not always the case that selection for treatment is based on the kind of threshold specified for the Hauer correction. For example, in many cases where a fixed budget is involved, it may be necessary to treat the 'n worst cases'. In Hauer's example, begin with the worst intersection, then the second worst, and so work down the line until the budget is exhausted. It is clear that something like Hauer's correction should work in this case, but unless the treatment programme stops exactly at a threshold number, we do not know what it should be.

Use of Statistics

The use of statistics in evaluation relies, by and large, on quite standard, accepted parts of statistical methodology that can be found in elementary textbooks. The question arises, therefore, as to why such well-accepted methods should be regarded with caution when applied to project evaluation.

A typical textbook example showing how to test for the difference between two means, might read as follows: 'An examination was given to 50 girls and 75 boys; the girls made an average grade of 76 with a standard deviation of 6, while the boys made an average grade of 82, with a standard deviation of 8' and go on to estimate the difference between true mean values. It might be tempting to translate this directly into a project evaluation by rephrasing the example in this way: 'Failure in 50 schools with classical teaching methods were compared with 75 schools having modern teaching methods, and the number of below standard results on the former averaged 76 (per year) with a standard deviation of 6, and on the latter 82 (per year) with a standard deviation of 8'. What is the basic difference between the two situations? The book example shows that there is indeed a significant difference in the two groups. But there are radically different aspects to the school evaluation example. First, the purpose of the experiment is to decide whether the new teaching method is (cost) effective, with a view to adopting it in the future (there is no thought of changing boys into girls to improve test scores). Second, there may be significant other factors, such as the quality of teachers which make the methods suspect as the 'cause' of the difference. In the textbook example, no discussion is given of the possible causes of the difference (e.g. age, education, motivation, etc.) and, even if such factors were uncovered by the school statistician, this fact would presumably not weaken the conclusion that 'for whatever reason, girls scored better on this test'. The corresponding conclusion that 'for whatever reason, one type of instruction is better' is not likely to be strong enough to influence future decisions, until such time as the 'whatever reasons' are fully understood. Finally, there may be trends which influence the evaluation case, but which have no significance in the boy-girl case.

In most evaluations, questions will come up about collecting data, storing it, analysing and manipulating it and forming conclusions from it. From one point of view these problems are simple. Virtually any elementary statistics textbook will give definitions and formulas that can be usefully employed. In particular, most texts give an inventory of various tests, showing the circumstances where they would best apply.

But there are some more difficult decisions to be made than this simple outline may suggest. For example, there is the question of which data will be needed, and how it can reasonably be obtained and at what cost. These are questions that the evaluation team should confront, if possible before the implementation begins.

We call this modelling, that is choosing the outline of the experiment. Should the data required be discrete (obtained by counting) or continuous (obtained by measuring)? We can count apples, or measure their weight. Contrary to the old slogan, statisticians have no problem with comparing apples and oranges. It can be done by many measures: weight, quantity, value, size, price are only a few.

Another consideration in formulating a model is finite vs. possibly infinite. If the experimental design involves the number of occurrences during a day, the model should be discrete and normally finite. But if the experiment required a variable representing the time to the first occurrence, it would be better to have a model, which was continuous, and possibly infinite (in this context, 'infinite' means only that it is not convenient to specify any fixed limit). Confusion can often arise from the wrong choice of finite/infinite and discrete/continuous.

Each of the possibilities corresponds to some statistical distribution model. For discrete and finite, it would probably be best to choose the binomial distribution. For discrete and infinite, the Poisson or negative binomial are often used. For continuous and finite, the beta distribution is better. Finally, for variables, which are thought to be continuous and infinite, there is the gamma distribution for quantities which are necessarily positive, and the ever-popular 'normal', Gaussian, or Laplace distribution for variables which are not always positive. Here is a classic example of what can happen with the wrong choice. In the early part of the Twentieth Century, the normal distribution was indeed the norm, so much so that other distributions were hardly considered. In his experiments with radioactive decay, Ernest Rutherford published a paper in which he despaired of fitting his data with the normal. The data was the time to decay, which, like many other times (time to the next telephone call, etc.) has quite a different distribution, namely the exponential one.

Each of the distributions carries with it several parameters: for example, mean, and standard deviation. Expressed in this way, the conclusion of an evaluation may well be based on a comparison between a sample and a modelled distribution. Such statistical inference (estimation and hypothesis testing) is well beyond the scope of this chapter; we only emphasise that whatever the method used, there is a better chance of success if a good model has been chosen. The key word is 'chosen'; that choice is the primary responsibility of the evaluation team.

When the model is chosen, and the test statistic, otherwise known as the measure of effectiveness, specified, then the technical or statistical part of the evaluation should be easy: just plug in the sample to the computer and wait for a red or green light to flash.

As we will see, the statistical part of the evaluation is by no means the whole job, and more difficult problems remain. Still, it is important to remember that a good, thoughtful beginning is important in most cases, and in some it is essential.

I do not propose to go into technical details, but would like to give one example where a clever set-up leads to simple evaluations. In a study of violent crime, it was desired to evaluate hospitals by the before/after number of admissions for violence. The idea was to see whether they had been improved by a certain intervention, perhaps a juvenile curfew.

Since the number of hospitals was large, and the number of admissions to any one was small, it was desired to pool the data by putting several hospitals together. But which ones should be grouped together? Some had large population catchments, others small; some had, we suppose, better facilities, some worse. Some had fewer admissions after the intervention, some more.

To balance the imponderable factors, it was decided to put together those hospitals with the same total number of admissions for violent crime before and after the intervention. Those with six, for example, could be categorised into seven types: (6 before, 0 after) (5 before, 1 after) (4 before, 2 after) and so forth until we come to (0 before, 6 after). The model would be binomial, where the theoretical proportions would be 1, 6, 15, 20, 15, 6, 1. Standard statistical methods could then be brought to bear to see whether the curfew had helped or hurt these localities. Also, by pooling those with the same totals, some measure of equalisation between hospitals would have been built into the model.

There are several advantages to this scheme: first, it is theoretically sound; second, the data is usually easy to obtain and easy to analyse; third it can be checked by choosing other categories of sites, say for 10 or 12 admissions; fourth, the method of selection often will amount to a random sample of all sites, since there is no bias in the selection for 'before' or 'after'. As usual, there is one conspicuous difficulty: there is no way of knowing whether any changes we discover were in fact caused by the curfew or by some other circumstances.

Another method that could be used in this case is called paired comparisons. Suppose there are 10 hospitals with one year admission counts as follows:

Hospital	Before	After	Reduction
1	81	76	5
2	52	60	- 8
3	87	85	2
4	70	58	12
5	86	91	- 5
6	77	75	2
7	90	82	8
8	63	64	- 1
9	85	79	6
10	83	88	- 5

To find a 98 per cent confidence interval for the true reduction in accidents, the first step is to calculate the mean and standard deviation of the observed reductions; this gives m=1.6 and s=6.38, which leads to a confidence interval (- 4.09, 7.29), and we can be 98 per cent sure that the true reduction lies in this interval. Inasmuch as the interval contains the value zero, it is impossible to say with assurance that the countermeasure has been effective in reducing accidents at these sites. However, the best estimate of the reduction is 1.6, so there appears to be some small effect.

Complex Modelling

Where there are several factors to be considered, the modelling can be remarkably complex, often involving expertise in higher mathematics. Typical of this situation are projects involving human behaviour, where the evaluation can be fragmented by age, income, location, etc. When successful, such a model will give useful information about the effect of the intervention on different classes, at different times or in different stages.

Time Series Analysis

An important difficulty with before-after analysis is the presence of trend in the data. One of the best known examples of this is the U.S. fatality rate per distance travelled. This has been systematically declining since the data was first recorded, about eighty years ago, from 20 fatalities per hundred million vehicles miles of travel to considerably less than one. Thus, a project devised to improve traffic safety, if measured by this particular parameter, is virtually guaranteed to 'succeed' if the sample size is large enough to smooth out minor blips. Since the trend is exponentially decreasing, a logarithmic transformation produces a nearly flat line. There is an interesting aspect to such data. The further we go along the time axis, the smaller the opportunity for further improvement. There is little doubt that the extremely expensive programmes of the nineties have had smaller effects than simple slogans, like 'Be Careful' did in the twenties.

Not all time series are this easy to describe. Some seem like random fluctuation; others have significant bumps. Many have periodic cycles, especially annual cycles, but also monthly or daily cycles. And, of course, virtually all have irregular minor ups and down. One purpose of time series analysis is to find a trend, if there is one, and remove the trend by a suitable mathematical transformation.

It has long been popular to 'evaluate' programmes by holding up for inspection a graph showing some change, over time, of the measure of effectiveness. This technique, based on nothing more substantial than intuition, can sometimes be quite persuasive. There are several famous examples: The introduction of compulsory seat belt wearing in the Australian state of Victoria, implementation of breathalyser legislation in Great Britain and the combination of oil panic, 55 mph speed limit and economic recession in the United States.

Whether or not such illustrations are persuasive depends to some extent on the skill or otherwise of the cartographer; 'how to lie with statistics'. Still less convincing are 'series' based on very few data points. Also, once again, nothing is shown about cause and effect. I have heard the flattening fatality rate per vehicle miles of travel be described as a consequence some ideal programme relating to suppression of alcohol, juveniles, carelessness, road rage, improved emergency care as well as increased seat belt wearing. It might be noted here that many professionals in the field believe that more important factors are demographic: an ageing driver population and increased urbanisation. Both of these are associated with lower travelling speeds. Clearly it would take a quite sophisticated analysis to disentangle all these influences. That might be another example of an evaluation not worth doing; it doesn't fit any of the criteria for evaluation: need to know, cost of knowing or chance of success.

When there has not been any intervention, we speak of trend analysis; when there has been an intervention, we use 'Box-Jenkins' or 'Box-Tiao' methods, both of which are rather complicated. There may also be a Hawthorne effect, requiring some elapse of time for the series to reach equilibrium.

The usual strategy for the analysis involves the following steps: model the long term trends present in the data series. Make forecasts for the post-intervention time periods. Base these forecasts on the pre-intervention data assuming a continuation of the long term trend, with no intervention effect. Compare the actual post-intervention data values with the values, which have been forecast. Judge whether the differences have a pattern. For example, if the actual values are consistently lower than the forecast values, the data can be taken to support the hypothesis that the intervention changed the level of the series.

Box-Jenkins Time Series Analysis (ARIMA) The second level of statistical techniques which is useful for implementing the basic strategy is Box-Jenkins time series analysis. Box-Jenkins models for time series data are quite powerful and flexible, but the model building process does require that a relatively large number of data points are available prior to the intervention. A popular rule of thumb states that at least 100 data values be used, although many successful analyses have been done using as few as 60 data points (even this number, for monthly data, would require five years' record). If fewer data points are available, the Box-Jenkins models should not be used.

Intervention Analysis for Time Series Models Box and Tiao (1975) give a method for using ARIMA models to measure and evaluate interventions. First the analyst must specify the general shape of the change caused by the intervention, classified as one of four types: (a) abrupt start with permanent duration (example: a speed limit change); (b) gradual start with permanent duration (example: improved safety standards for new vehicles); (c) abrupt start with temporary duration (example: a public information campaign); (d) gradual start with temporary duration (example: a selective enforcement campaign).

Control Group

In the preceding discussion, several difficulties with the interpretation of before-after comparisons have been found. An alternative is to conduct a controlled experiment. During the same time period, a random selection of sites, the 'experimental group' is treated, while another random selection of sites the 'control group' is left untreated. If the experiment is properly designed the control group and the experimental group will be subject to the same exogenous influences. Thus, the experiment should measure only the intervention, so that the principal difficulty with before-after design is avoided.

But here again, evaluation of social projects presents difficulties. In studying agriculture, for example, it is perfectly feasible for the experimentalist to vary soils, fertilisers, seeds, etc. at his pleasure, and to analyse the resulting yields. In social projects, on the other hand, it is not usually possible to control different variables as the evaluator would wish, merely for the sake of experimentation. Often the intervention is embedded in law, or in custom. This is particularly true for the so-called gold standard of evaluation, randomised sampling. When we consider the difficulties and especially the costs of randomised experimentation, it is understandable that evaluations of social projects are so often by necessity before-and-after studies.

In a study of posted speed limits in Sweden about 40 years ago, an elaborately designed experiment, although it produced mountains of data, failed to correlate strongly any particular speed limit with any particular accident risk. This was certainly not because there was no relation, but simply because the relation was too feeble to be easily be teased out from the data. However, all was not lost; by using the data in a carefully selected manner, the public and the political authorities was convinced that a speed limit was a good thing.

Some assignments to control groups are difficult or even impossible. When an experiment was designed to evaluate a spectrum of punishments for drunk drivers, some judges upset the scheme by refusing assignment of particularly egregious offenders to the low penalty option.

Even if possible, random assignment may violate ethical standards or those acceptable to the people affected. Public opinion will very likely oppose treating individuals or groups, or schools differently based only on the flip of a coin. In studying youth curfews, it is hardly possible to impose it only to those whose names begin with A to L, but not in those with M to Z names.

One rather appealing variant in design is the so-called 'week on/week off' where services or sanctions are applied only at a randomly selected number of days or weeks. In some circumstances, the approaches true random assignment, and it is usually easier to implement and thus appealing to the evaluation staff.

The handbook produced by the (Washington, D.C.) Urban Institute contains this caution: 'Great care needs to be taken to ensure that no members of the control group are denied any essential services they would otherwise have, that the benefits to participants and the community are carefully explained, and that the program staff and participants understand and support the research' (Harrell, A. *et al.* (undated)).

Programme Evaluation

We now turn to a more difficult subject, namely a discussion of methods for evaluating an entire programme consisting of many projects, some related to one another – for example conducted in different years or different places – and others perhaps imported from similar work in other jurisdictions or even from other countries. This process is called meta-analysis. The discussion given here is based on the work of Rune Elvik.

Meta-analysis denotes quantitative techniques for summarising the results of a set of projects made to evaluate the effects of similar interventions. The starting point is the fact that evaluation research is often controversial. Controversies tend to arise when the results are unexpected or counterintuitive. Examples of counterintuitive results from road safety research in Norway include the finding that marked pedestrian crossing facilities increase the number of accidents and that skid training of car drivers increases the number of accidents. Results like these are met with disbelief. A relevant question then becomes 'When can we trust evaluation studies?'. What characterises a good evaluation study, and what characterises a poor evaluation study?

Some people might be inclined to say that it is impossible to identify good and bad evaluation research. In the final analysis, it all boils down to whether we like the results of a study or not. Here it is argued that comparatively objective criteria of good evaluation research can be developed. The term 'comparatively objective' implies that the criteria of good evaluation research are: (a) stated in sufficiently clear terms to rule out highly diverging interpretations; (b) based on methodological principles and rules that are very widely (but perhaps not universally) supported by researchers, and not at least, (c) independent of the results of other studies, and therefore also independent of whether we 'like' or 'dislike' these results.

In Elvik's work, criteria of good evaluation studies have been developed within the framework of the validity system proposed by Cook and Campbell (1979). In this framework, the validity of a study or set of studies is defined as approximation to the truth. The more and stronger reasons we have for believing that a study or set of studies comes close to the truth, the higher is the validity of that study or set of studies.

Statistical conclusion validity refers to the numerical accuracy, reliability and representativeness of the results of a study or set of studies. The criteria are the sampling technique, sample size, measurement reliability, the presence of systematic errors in data, choice of technique of analysis, commensurability of the dependent variables in a set of studies, publication bias, the shape of the distribution of a set of results, particularly in terms of modality, skewness and outlier bias and finally the robustness of the mean result of a set of studies with respect to how it is estimated.

Theoretical validity denotes the extent to which a study has an explicit theoretical basis that provides an explanation of the findings of the study. The criteria are whether an explicit theoretical basis has been developed for a study; the possibility of giving adequate operational definitions of theoretical concepts used; whether the theory on which a study is based can contribute to explaining the findings of the study or whether the theory on which a study is based is supported by the findings of the study.

Internal validity refers to the possibility of inferring a causal relationship between the measure that is being evaluated and the dependent variables this measure is intended to influence. It should be possible to identify a causal mechanism that explains why the cause produces the effect. The relationship between cause and effect should be reproduced in several studies, preferably made in different contexts. If there is sufficient variation in both cause and effect, there should be a dose-response relationship (that quantitative changes in one variable are reflected proportionally in the other) between cause and effect and finally, if an effect is believed to exist only in certain group, it should be found only in that group and not outside it (specificity of effect).

External validity refers to the possibility of generalising the results of a set of studies to other contexts and settings than those in which each of the studies was made. This kind of generalisation is often desirable in evaluation research. One wants to know, for example, if the results of studies made in countries A, B and C apply to country D as well. Generalising across countries in this manner is common in evaluation research, since not every country can do its own research in every subject. The task facing those who want to extract the best-established knowledge from this research is, simply put, to sort out the good studies from the bad ones. Meta-analysis can help in accomplishing this task, but it can never capture all relevant considerations in assessing study quality. There are aspects of study quality that do not lend themselves to numerical coding and cannot be brought within the framework of meta-analysis. It is nevertheless obvious that meta-analysis can be widely applied to evaluation research.

Multiple Goals

Suppose the intervention is designed to achieve not one, but several goals. We may wish to reduce pollution from 8 to 7, speed traffic from 5 to 6, increase public approval from 4 to 5, etc. More and more projects are like this, and present difficulties for the evaluation team. For example, the USDOT handbook (Casey and Collura, 1994) states that one project is intended to (a) enhance the ability of public transportation to satisfy customer needs and (b) to contribute to the broader community goals. Each of these is broken down into a number of 'objectives'. These examples are typical of the entire list of 21 goals: 'reduce transit system costs', 'improve communication with users having disabilities' and 'assist in achieving air quality goals'.

Even if each of these goals can be expressed in terms of a single parameter, the maximising each of the 21 variables is beyond any analytic capability. The best that can be done is to estimate the success of the project in each (or to measure the cost of each), and leave it up to the management to choose how to interpret the result. It would not be surprising if the choice were made among those which are approved by the project management, and let the others be downplayed or omitted entirely.

Such a simple and basic example shows that the evaluation of a complex project must, in the end, involve opinion. Furthermore, it seems likely that the prevailing opinion will be that of the project management, rather than that of the evaluation team. The purpose of the final report of the evaluators in the handbook cited above is to synthesise the findings relative to each of the program objectives and other relevant objectives/issues. The evaluation team can perhaps find the effect of a single intervention on each of the goals specified by the administrator, if these are clearly stated. But relevant objectives/issues is undeniably vague, and it is hard to see how the evaluation team, absent some guidance, can fulfil the requirement without a substantial injection of its own opinions.

If the intervention has some well defined, specific criteria of success, it is reasonable for the evaluation to assess how well each criterion is being met, but not to guess which is more important.

The Cost of the Intervention

This is another difficult problem in some evaluations, especially when the administrator wishes to have a positive result.

The project managers will be able to furnish estimates of cost, which are more or less accurate, and more or less complete. The evaluation team should consider carefully the information, not only from the point of view of accuracy, but also to see if it does indeed represent the true cost of the project, not only to the contractor, but also to other individuals and to society as a whole.

The U.S. Federal Highway Administration evaluation manual calculated the cost of a STOP sign at $75 for installation, $5 per year for maintenance, with a salvage value of $10. On this basis, using the cost of life and injury, it was clear that virtually any STOP sign in a densely populated area was cost-effective. However, the calculation omits the cost in vehicle wear and fuel consumption of the stopping and starting of vehicles obeying the sign. This would be particularly important in a country where both vehicles and fuel represented investment in hard currency. Shall the evaluator recommend that a low-income country be sprinkled with STOP signs, as a life saving measure? But the cost of vehicle wear and fuel consumption is doled out in small increments, not to the public purse, but to the user of the vehicle. Shall the evaluator engage in 'sub-optimisation' as the administrator clearly wishes? Here we have an ethical difficulty.

Evaluation Synthesis

The really difficult part for the evaluation team is evaluation synthesis. In forming a coherent picture of the effects of the intervention, just as in the original evaluation design, the expertise of the evaluation team will be needed. The primary objective of evaluation synthesis is to check, from the evidence, various aspects of the data gathering process, the statistical analysis and formation of conclusions. This may involve such mundane tasks as looking over worksheets to see that no items have been omitted.

In an accident monitoring project, it was discovered that the staff failed to make records during an important public holiday, because they were 'off duty' for the celebrations. As this example illustrates, the first consideration must be the administrative evaluation. This means some careful checking that the entire intervention took place exactly as intended, with no exceptions, no loopholes and no weakness.

Second, if it appears that there is a correlation between the intervention and the measures of effectiveness, a really determined effort must be made to see if the effect is really cause and effect. There are many candidates for alternate causes. First the state of the economy, which is very important in influencing travel behaviour and traffic safety, and I daresay in other social programmes as well.

In the evaluation of the 1974 speed limit, reported earlier, the speeds also dropped on roads that already had speed limits below 55. That was explained away as 'merely halo effect'. But, drownings also fell substantially perhaps owing to the severe recession, and there was a nation-wide truckers strike (a recession is often correlated with increased safety, this is normally attributed to the cancellation of holiday travel, which has highest vehicle occupancy, and thus greatest risk of fatality in any collision).

A particular side effect of an accident project, which may occur in some cases, has been called 'accident migration'. This means that accidents prevented at one place or time, or of one type, may 'migrate' to another place, time or type. The existence of migration is most plausible with respect to accident type. In a project of intersection signalisation, although side impacts may be reduced or eliminated, it is logical to think that rear end collisions may be increased. Boyle and Wright (1984) find that there is geographical migration as well. In a study of 133 London sites, they found a 22 per cent decrease in accidents at the sites, combined with a 10 per cent increase in the immediately surrounding sites. In a subsequent paper some effect of regression to the mean is acknowledged.

It is a question for the evaluation team, to what extent to include possible secondary costs and secondary effects in the synthesised evaluation. In addition, there can be questions of pitfalls in the original measures of effectiveness. The point has been made that guardrails can be expected to increase accident frequency while decreasing severity. Similarly, provision of emergency medical services and hospitals can be expected to increase accident severity, if severity is measured on a scale such as the following: (a) no injury (b) injured but not hospitalised (c) hospitalised (d) fatal. Joksch (1975) points out that 'a safety feature eliminating minor injuries might eliminate such crashes entirely from the record, retaining only more severe injuries, so that injury severity appears to be increasing'.

Failing to identify exactly what the other issues may be, and failing to specify some method for synthesising the findings, leaves the evaluation team with no exactly specified task. As the following example illustrates, a convenient substitute for exact information, which avoids reliance on the opinion of the project management as well as of the evaluation team, is to measure public opinion.

We would hesitate to use public opinion to determine the distance to Mars, or even to estimate the population of China. But projects having multiple goals are frequently dependent on and sensitive to political acceptance, however unscientific that may be. A key element must be to find knowledgeable opinion.

Transportation research provides a good example where informed opinion is a possibility. Then, the task of the evaluation team is greatly simplified. Instead of asking 'is it so?' one need ask only 'is it believed to be so?'. Using such a method may lead to bad policy decisions, but it may satisfy the evaluation team, and will almost surely satisfy the project administrators.

As an example, I would like to mention briefly a project in California. This project throws some light on one way to integrate public opinion with stated objectives (see Golob, 1999). There is a 13km stretch of freeway ('The I-15') near San Diego, which has experienced some congestion. The freeway contains a lane open to high occupancy vehicles only (in California high occupancy means at least two persons per car). This HOV lane was not fully utilised, and a scheme ('FasTrak') was proposed to let single occupancy vehicles use the HOV lane, for a variable fee. The fee would be adjusted for most acceptable flow (to traffic engineers or municipal authorities.) But instead of defining 'most acceptable' in numerical terms, it was defined by reference to users' opinions.

The FasTrak project was to be evaluated in terms of 'congestion relief, land use, business impacts, and media coverage' Golob's paper deals with the public opinion of these goals, and is based on the analysis of a panel. A panel has the advantage that it can accurately track the opinion of users over time, before, during and after the project, and presumably provide guidance in deciding on similar projects, and in designing experiments to evaluate them.

With so many variables (age, household income, commute distance, gender, etc.) some fairly complex experimental design was required. Turning now to a sample of the outcomes, it was found that the perceived time saved did not vary substantially over several years of experimentation, with some variability in the conditions: it was a bit over sixteen minutes in one direction and about twenty-two minutes in the other.

There was a discrepancy between the tabulated results (mostly time savings and satisfaction) with the stated purposes of the project: (a) to maximise use of the I-15 Express Lanes (b) test whether allowing solo drivers to use the Express Lanes excess capacity can help relieve congestion on the main lanes (c) improve air quality (d) fund new transit and carpool improvements in the I-15 corridor, and (e) test using pricing to set tolls.

The modelling and analysis could well be a model for the best U.S. practice in 2000, but with regard to the evaluation standards there is one significant gap: addressing the stated goals of the implementation. It may well be that the goals of the project were devised a posterieri, and can be adjusted to the results obtained.

References

Andersson, B.J., Erlander, S. and Gustavsson, J. (1963), *Tillfällig Hastighetsbegränsning I motortrafiken*, Statens Offentliga Utredningar, Stockholm.

Boyle, A.J. and Wright, C.C. (1984) 'Accident 'Migration' After Remedial Treatment at Accident Blackspots', *Traffic Engineering & Control*, Vol. 25, No. 5, pp. 260-267.

Casey, R.F. and Collura, J. (1994), *Advanced Public Transportation Systems: Evaluation Guidelines*, U.S. Department of Transportation, Washington, D.C. (DOT-T-94-10).

Cook, T.D. and Campbell, D.T. (1979), *Quasi-Experimentation : Design & Analysis Issues for Field Settings*, Houghton Mifflin, Boston.

Council, F.M. *et al.* (1980), *Accident Research Manual: Final Report*, University of North Carolina, Highway Safety Research Center, Federal Highway Administration Report, No. FHWA/RD-80/016, Chapel Hill.

Elvik, R. (1999), *Assessing the Validity of Evaluation Research by Means of Meta-Analysis*, Institute of Transport Economics (Report 430), ISBN 82-480-0091-5, ISSN 0802-0175, Oslo.

Golob, J.M. and Golob, T.F. (2000), 'Studying Road Pricing Policy with Panel Data Analysis: The San Diego I-15 HOT Lanes', presented at the *Ninth Conference of the International Association for Travel Behaviour Research*, Gold Coast, Queensland, Australia, 2-7 July 2000.

Golob, T.F. (2001), 'Joint Models of Attitudes and Behavior in Evaluation of the San Diego I-15 Congestion Pricing Project', *Transportation Research* (forthcoming in Part A, Vol. 35).

Haight, F.A. (1985), 'Conceptual Pitfalls in Traffic Safety Evaluation', *Evaluation 85*, Colloque International sur l'Evaluation des mesures locales de securite Routiere, O.N.S.E.R. 94114 Arceuil Cedex.

Harrell, A., Burt, M., Hatry, H., Rossman, S., Roth, J. and Sabol, W. (undated), *Evaluation Strategies for Human Services Programs; A Guide for Policymakers and Providers*, The Urban Institute, Washington, D. C.

Hatry, H.P. (1980), 'Pitfalls of Evaluation', *Pitfalls of Analysis*, International Series on Applied Systems Analysis, Vol. 8, John Wiley and Sons, New York.

Hauer, E. (1980), 'Bias by Selection: Overestimation of the Effectiveness of Safety Countermeasures caused by the Process of Section for Treatment', *Accident Analysis and Prevention*, Vol. 12, No. 2, pp. 113-118.

Heckard, R.F., Pachuta, J.A. and Haight, F.A. (1976), *Safety Aspects of the National 55 mph Speed Limit*, U.S. Department of Transportation, Federal Highway Administration, FHWA-RD-76-191.

Higgins, T. J. and Johnson, W. L. (1999), 'Evaluating Transportation Programs: Neglected Principles', *Transportation*, Vol. 26, pp.323-336.

Joksch, H. (1975), 'A Critical Appraisal of the Applicability of Benefit-Cost Analysis to Highway Traffic Safety', *Accident Analysis and Prevention*, Vol. 7, No. 2, pp. 133-154.

Majone, G. and Quade, E. S. (eds) (1980), 'Pitfalls of Analysis', *International Series on Applied Systems Analysis*, viii, 213, Chichester, New York.

Public Technology, Inc. (undated), *Program Evaluation and Analysis: A Technical Guide for State and Local Governments*, U.S. Department of Housing and Urban Development, Office of Policy Development and Research, Washington, D. C.

Rossi, P. H. (1977), 'Boobytraps and Pitfalls in the Evaluation of Social Action Programs', in Caro, F. G. (ed.), *Readings in Evaluation Research*, Russell Sage Foundation, New York.

Viatek Consulting Engineers (1980), *Road Safety Improvement Project. Study Report*, Ministry of Transport and Communications, Kenya with Ministry for Foreign Affairs, pp. 108 and 13 appendices, Finland.

Waller, J. (1985), *Injury Control: A Guide to the Causes and Prevention of Trauma*, Lexington Books, Lexington.

7 Transport Evaluation Methods: From Cost-Benefit Analysis to Multicriteria Analysis and the Decision Framework

MICHEL BEUTHE

Introduction[1]

Over the last decade, the European Commission has commissioned a series of research programmes aimed at developing practical methodologies that allow the assessment of European large-scale transport projects. With regard to assessment methodology per se, the main effort started with the EURET Concerted Action, the main task of which was to review the project appraisal practices applied in the Union's Member States for the different transport modes – road, railway and inland waterway – and to recommend appropriate steps towards the establishment of a co-ordinated European assessment methodology. EURET was followed by the APAS programme which continued this methodological effort. In the Fourth Framework Research Programme, the EUNET/SASI consortium – composed largely of research teams that had participated in EURET and APAS – initiated some experimental implementation of methodologies. EUNET made a remarkable effort of gathering and deriving comparable data which could be used in the assessment of international projects inside and outside the Union, it proposed operational solutions among competing methodologies, and it applied these to demonstration case studies.

EUNET, as much as its predecessors APAS and EURET, was mainly concerned with the efficiency aspect of transport initiatives. Their strategic policy relevance and implications were tackled by two other projects, namely TENASSESS and CODE-TEN which together developed a method for strategic transport assessment and applied this to several corridor investment programmes.

In this paper I reflect on the choices and options made in applied research for the purpose of developing an operational assessment methodology. To do this I rely on the knowledge and experience gained through marginal participation in some of the aforementioned projects or programmes. I should make it clear at the outset that my purpose in this paper is not to evaluate these research projects. Rather, I use them as references to exemplify specific issues of importance for transport evaluation.

The first part of the paper reviews the progressive evolution of project appraisal and discusses the question of whether cost-benefit analysis (CBA) – as a method of valuing project outcomes in money values rather than in utilities – should not indeed be our first choice of an evaluation methodology when budget constraints are relevant. To answer this question I review in the first section the basic principles of cost-benefit analysis and compare these to present practices. Next, I examine how multicriteria analysis (MCA) is used in the context of the assessment of transport initiatives. Finally, in the third section I discuss the evolution from decision-making analysis to a decision-support approach. The latter aims at providing a framework for group decision making and negotiation rather than a selection of the best projects.

The second part (fourth section) examines whether the uncertainty which characterises a project outcome should not receive more attention than what it currently does in appraisal practice. I review a number of methods which have been proposed to handle uncertainty, and, lastly, propose a method to transform utilities in equivalent money values which allows an estimation of risk premiums.

The final part (fifth section) discusses the problems presented by wider range criteria like spatial accessibility, social cohesion, strategic economic development, employment and environmental effects which are introduced in several assessment methodologies, as in EUNET and CODE-TEN.

The conclusions will recap the main line of the argument followed in the paper, focusing on money valuation, assessment of uncertainty and the problems of wider ranging criteria.

Monetary Assessment of Projects

Cost-Benefit Analysis with Market Values?

Within conventional economic analysis, when a consumer decides to buy a good, it is because the satisfaction or the utility derived from it is at least as great as the utility he or she would obtain from saving the same amount of money for later use or from spending it on some other item. That good is worth at least the price the consumer is paying. As all people behave in the same way, they will buy more of the good so long as the utility of an additional unit is – for some of them – higher or at least equal to the utility of the money they have to pay. This marginal assessing process by consumers determines the quantity they buy. As at any point in time a unit of the good is just worth its price, the latter can be taken as a measure of the marginal utility obtained by the consumers.

In similar vein, the producers of the good are ready to supply more of it so long as the price they obtain covers the cost of the factors needed for the additional production (at least), up to a quantity such that the last unit of output has a cost equal to the price. Hence, the price is also a measure of the marginal cost of production. This is the case in a situation of perfect competition with many producers, when market organisation and regulation do not deflect prices from their proper economic role and level. In such a situation, the equality between price and marginal cost guarantees that productive resources are allocated to meet the consumers' preferences for a given distribution of incomes.

A similar interpretation can be given to prices of inputs like natural and human resources, which are bought by firms and consumers according to their marginal productivity and marginal utility. Hence, in a situation of perfect competition market prices and money provide a convenient and uniform measure of the marginal worth of goods and services for consumers as well as for producers. Moreover, in matters of public investment, the budget involved in a project is also a measure of the utility of forgone consumption. It is on this basis that the Cost-Benefit Analysis (CBA) of private projects – conducted in terms of monetary market gains and costs – was extended to public projects which must be assessed for their contribution to population welfare. In principle, the political decision-maker should accept a project implementation if its general public utility is at least as great as the utility of spending public funds on other projects and/or the utility of the money involved for taxpayers. By exploiting the correspondence between money and utility as outlined above, cost-benefit analysis can value the negative and positive impacts of a project in monetary terms and thus recommend the implementation of projects with a positive net (present) value. Thus defining 'added value' allows cost-benefit analysts to claim that they are in a position to identify those projects which contribute to an increase of the population welfare.

However attractive this approach may be, it meets a number of difficulties which cannot all be completely resolved. First of all, most prices are not set in markets meeting the conditions of pure competition. In a monopoly situation, for instance, the price of a commodity may very well be higher than its marginal cost. If this commodity is an important productive input in a project, this means that the project cost as computed in market value is overestimated and its net value underestimated. Hence, the project could be rejected even though it should have been implemented on the basis of a correct estimation of its economic resource cost: not its apparent cost, the monopoly price, but its real cost, its so-called shadow price, which is the production cost of that input. More generally, this means that monetary valuations of impacts based on market prices cannot be made without paying attention to market conditions and organisation. For that reason a substantial part of the literature on cost-benefit analysis is devoted to careful analyses of appropriate prices and values to apply in project assessment.

Beyond this problem of non-competitive markets it is clear that not all social impacts can be valued from market prices since some outputs or inputs do not have any market. There is a market price for working time, but it can be argued that there is no price for leisure time, while there is definitely not a market price for clean air, life saved or the preservation of environment. To solve that type of problem and encompass social impacts and externalities in what should be called a social cost-benefit analysis (SCBA), a number of approaches have been proposed. These are based on the observation of choices made by people. For instance, the higher wage obtained for a dangerous work may be used to derive an estimation of the value given to one's life. Travellers choices regressed on the multivariate attributes of different travelling alternatives including cost may provide an estimation of the value attached to time saved, comfort or frequency. Distance travelled and its associated cost to visit a protected park may give an indication of its worth for the visitors. Needless to say that such methods, the so-called revealed preferences methods, give rise to vigorous debates concerning possible estimation biases.

An alternative to analysing observed choices and preferences is to interview people on the choices they would make between well-specified, fictitious albeit realistic alternatives. Their stated preferences set in relation to the circumstances and consequences of their choice may then be used to estimate the value they attribute to life, exposition to diseases, and quality attributes. This method may even allow making estimates of the option value imputed to the existence of a facility or service, like public transport, even when there is no planned intention to use it. Obviously, for reliable estimates the questionnaire design demands careful consideration as biases can be easily introduced when the questions are not well understood or tend to suggest particular responses. Furthermore, keeping in line with the cost-benefit analysis search for market values, the questions raised should be set in the context of a realistic market as far as possible (Layard and Glaister, 1994; Nash and Mackie, 1990; Schmid, 1989; Stiglitz, 1988).

When an impact cannot be considered marginal for every citizen, the market prices – observed or estimated – may not suffice to assess an impact value since prices are concepts of marginal value. In such cases, it is necessary to estimate the 'consumer surplus' which corresponds to the total willingness to pay. That requires an estimation of the compensated demand curve, a difficult task that is dealt with by a rough approximation in most cases.

Another difficulty concerns the discount rate that applies. Particularly with regard to long-term projects, the choice of the discount rate can strongly affect the assessment result. Should one use the borrowing rate of the State, the rate paid by consumers or the one of corporations? Should one base the rate on the shadow capital cost of public investment under budget constraints? The solution may depend on the circumstances of the projects, but there is certainly no clear consensus on this issue. In the end, this is a matter of judgement with respect to the overall circumstances of the project and the objectives set by the State concerning inter-generational justice. Low discount rates reflect a strong concern with inter-generational equity and/or with negative environmental effects (hence sustainability) as opposed to financial considerations. Thus, in practice, the setting of the discount rate may be somewhat arbitrary and result from a normative public investment policy (Stiglitz, 1994).

At this point, it is worth mentioning the problem affecting appraisal of international projects where impacts are spread in different countries and over travelling people of different nationalities. This is a very difficult problem to handle in practice (Nellthorp *et al.*, EUNET Deliverable 9,

1999). An acceptable pragmatic solution is to use 'European values' computed as weighted averages. The problem with this solution – one should be aware – is that it in effect replaces market values by some kind of European value norms, one more step away from the philosophy underlying cost-benefit analysis.

Another difficulty relates to the unequal income distribution among citizens and the variation in tastes and preferences which is partly but not fully conditional on income. The marginal utility of income, i.e. the income's worth for the consumer, cannot be equal for all citizens. The consequence of this is that even though market monetary units can be used for measuring a particular consumer's utility, the values obtained for different individuals cannot be meaningfully compared. Were we to define collective welfare as the sum of individual utilities – a strong hypothesis – we could not simply add the corresponding monetary values to measure this collective welfare. Leaving aside the issue of different tastes and preferences as a minor one, it seems then that only a weighted addition taking into account the relative worth of income for different categories of consumers would make sense. Unhappily, this is not an easy matter, because the impacts need to be computed per category of incomes, and because the appropriate weights expressing the relation between marginal utilities and income cannot really be ascertained from empirical evidence.

However ingenuous and careful methods for deriving the money values of social impacts might be, there remain some important impact categories which cannot be estimated by reference to market situations and prices. In these cases an (expert) judgement of some kind is unavoidable. Impacts on different categories of income would have to be weighted differently, but the weights expressing the relation between utilities and income are highly speculative. In practice they result from an informed value judgement by the analyst or by the public authorities commissioning the study. A good example is provided by the Spanish rail evaluation manual where equity weights are given to three groups of income (Ministry of Public Works, 1996). Realistically such weights should be viewed as the outcome of a political process rather than of economic analysis. Likewise, the regional distribution of impacts may be an important consideration for a public decision-maker: for example, the benefits of a transport investment decision may accrue to the region where the former is being implemented while negative network effects are born elsewhere. It is hard to see how the determination of regional relative weights for such a policy assessment could be made outside the political process. As a matter of fact, regardless of the (weak) factual information which could be put forward, the weights given to impact distributions are actually decided by the responsible political authorities.

Similar remarks could be made on decision criteria like impacts on employment, network accessibility, economic autonomy or prestige, land use, and regional development. The theoretical understanding or empirical evidence are not satisfactory in these cases either. Hence, the only recourse is to weigh these criteria according to political willingness.

Another factor is the risk involved in a project, i.e. the risk of errors in cost and impact estimation, of technical failure or accident. This difficult factor is only too easily set aside for the reason that the consequences of a project are spread out thinly over the whole population (Arrow and Lind, 1970). This may be the case for small projects, particularly for the financial cost. However, there are huge projects with a substantial impact on the whole economy of a country; in this case, negative effects would lead to cumulative negative impacts on other activities. There are also projects with negative effects concentrated on only part of the population, as well as projects with environmental effects which are not divided and spread over the population as they bear on everyone with the same intensity (Fisher, 1974). In the current practice of social cost-benefit analysis this problem is addressed through a sensitivity analysis of assessment results with different possible values and parameters, leaving the decision-maker to judge whether the project is worth implementing despite the uncertainty characterising its outcome. However the sensitivity analysis is hardly ever based on any real assessment of the risks involved in the project or on the relative reliability of estimates introduced in the modelling. Naturally, it may be too much asking from an engineer or an economic analyst to raise doubts about their own design and estimates. Furthermore, appraising a project is a difficult and time-consuming task. Time is always a bit short for fully completing the task, while risk analysis – in any event a difficult task – comes at the end of the process. A possible avenue for improving this somewhat unsatisfactory state of affairs will be discussed in a later section.

Monetary Assessment with Multicriteria Analysis?

Social cost-benefit analysis cannot handle entirely the assessment of public projects that have impacts with no direct monetary valuation. The natural solution is then to pursue the assessment analysis with a multicriteria methodology (MCA). The latter is able to integrate various criteria measured in different units through an aggregating function which represents the decision-maker's preferences.

The passage from (social) cost-benefit analysis to multicriteria analysis is not a trivial one. In (social) cost-benefit analysis the decision-maker accepts as a matter of principle that market values, as determined by the preferences of consumers and the cost of production, delineate the reference for valuing a project. In contrast, the weighting of criteria in multicriteria analysis is based on the public decision-maker's assessment of the relative importance of the different criteria. The net present value as computed by (social) cost-benefit analysis can of course be taken as one criterion, yet its weight in the assessment is determined by the decision-maker. It is anticipated that the decision-maker represents truthfully the population's preferences (or welfare) by reason of the power of representation granted to him or her through voting, and hence that this is an efficient decision procedure. Nevertheless the step from (social) cost-benefit analysis to multicriteria analysis exposes more opportunities for easy arbitrary assessments, a danger which should always be kept in mind.

The extension of project assessment to include additional qualitative or political criteria, whilst necessary, may be the source of problems of double counting effects. For instance, an explicit goal of the European transport policy is to promote mobility within the Union. This could be measured through an accessibility indicator based on the cost or time for travelling a certain distance. However, such an impact may de facto be present on the (social) cost-benefit side of the analysis through the inclusion of time and cost savings in the estimation of the project's net benefits. Hence, if such a criterion is included at all in a multicriteria analysis, its meaning must be carefully and explicitly defined with regard to what has already been included in the (social) cost-benefit analysis. A similar problem occurs with regard to the measurement of employment effects, where the direct impact of a project on employment is already included in the (social) cost-benefit analysis through the value imputed to labour in the computation of costs. Again, a careful and coherent definition of this impact is necessary in order not to inflate the value of a project. An alternative solution is to separate (social) cost-benefit and multicriteria analyses, and include in the latter mainly the criteria that cannot be easily assessed in monetary terms. The implication of such a solution is that the idea of a final unique assessment value is thus abandoned. This is one type of solution considered by the EUNET research team.

Another serious problem is the definition of impacts and criteria if coherent comparisons are to be made between projects. If impacts are scored, as it is often done, on a scale from 0 to 100, great care is needed to set the domain of variation of each impact in a common unit of measurement prior to attributing a score. The decision-maker must remain well aware of these scales when setting the weights of the multicriteria function which will allow the aggregation of the scores. These weights should express the importance given by the decision-maker to the real consequences of projects. One may fear that the setting of an arbitrary common score scale will create some confusion in this respect, particularly when the scoring of impacts and the weighting of criteria are somewhat dissociated.[2] Actually, the weights must be taken as specific to each set of projects under analysis since their consequences and their range may be different from one set of projects to another.

Important furthermore to remember is that even if projects can be usefully ranked through a preference function, the latter cannot assess their worth with regard to the budgets involved. In contrast, working with market values (social) cost-benefit analysis can compute net present values and rates of return, thus providing rules for rejecting those projects not worth their budget. A workable solution to this difficulty involves the transformation in monetary units of the preference function using the trade-off rates implicit in the preference function.

Given the presence of a budget constraint, the choice made between different alternatives on the basis of monetary gains or costs implies de facto the monetary valuation of non-market criteria like safety, time savings, employment, industrial policy, etc. A budget constraint imposes so-to-speak an implicit price on everything, an opportunity cost or shadow price that translates the importance given by the decision-maker to the desired objectives, services or results. Since there is always an opportunity cost in this context, even though there is not a market price, we think it is preferable to use money values as a yardstick for assessing projects and their impacts. Money values are better understood by everyone, citizens as well as decision-makers. They maintain a direct link with market values based on consumers' preferences and production costs; they provide a readily understandable unit of measure for comparing impacts so that checking the coherence of the results is made easier; moreover, they allow the direct comparison with budgetary costs (Beuthe *et al.*, 2000). As suggested above (social) cost-benefit analysis is itself mixing consumer and political willingness to pay. Why not go a step further with multicriteria analysis whenever budget constraints – the usual case – are present? This recommendation and how to implement it will be further discussed in the section entitled 'Multicriteria Analysis and Uncertainty'.

The Fiction of a Single Decision-Maker: A Decision Framework

Another problem in contemporary transport evaluation is the fiction of the single and unique decision-maker. At the end of the political decision process, there may very well be one person who cuts all the remaining knots of opposite views among concerned people and their representatives, but in doing so, he or she is most likely choosing a feasible solution based on a compromise between different parties. Hence, however sophisticated, multicriteria analysis techniques of valuing projects can only be viewed as a way to process information and to identify relevant issues for the several decision-makers or the committee in charge. Multicriteria analysis cannot be used for ranking projects according to the preferences of one decision-maker. Rather its role is reduced to ranking projects under various hypothetical sets of weights (sensitivity analysis), leaving the task of choosing one particular set or a combination of them to the decision-makers. This means that beyond both (social) cost-benefit and multicriteria analyses there is a principal need for an analysis of group decision-making and of the role of expert advice in this process. This problem exists in every country but it is magnified at the European level where bargaining among Member States plays a major role, even if otherwise presented.

The recognition of the decision-making process as a factor of significance in reaching a decision on any project or initiative – in transport as in any other policy sector – has inspired proposals in favour of a different terminology in assessment and evaluation: one is today more likely to talk about a framework of analysis for decision-making rather than an appraisal methodology in the usual sense. The analyst recognises the fact that he or she cannot expect to integrate in a simple way all factors that will determine the final decision, particularly not those political elements, which enter into the dynamics of a group negotiation. As a consequence, the analyst aims only at providing as much information as possible on the different aspects of the concerned projects: a (social) cost-benefit analysis based on the impacts which are usually measured in monetary terms; some kind of a multicriteria analysis where additional impacts are introduced; a financial analysis focusing on factors which impact on the budgetary flows, like tax revenue; and information on particular items which are singled out for the decision-maker's attention. There is no effort made to identify any decision-maker's preference system, as would be done in a real multicriteria analysis, since such an attempt would be doomed: the decision makers are many, they will somehow decide upon consultation with each other, and, in the meantime, they do not wish to reveal openly their objective preferences and would not readily submit themselves to any kind of lengthy questionnaire. Hence, the analyst can only postulate some arbitrary set of weights purported to represent specific preference attitudes and proceed on that basis to some informative sensitivity analysis. The appraisal methodology thus tends to become a decision support system (DSS) in the line of what is proposed by Shakun (1996) and Hämäläinen and Pöyhönen (1996), among others, in order to prepare and support negotiation and group decision-making.

A good example of this evolution is the EUNET methodology. The EUNET research programme had as main objective the development of a comprehensive method for modelling and assessing the socio-economic impacts of strategic transport initiatives. This meant devising methods for providing input data and a methodology for the measurement and valuation of socio-economic effects in order 'to support the decision maker in assessing the merit of different projects' (Grant-Muller *et al.*, EUNET Deliverable 16, 1999, p.5).

In EUNET, a (social) cost-benefit analysis module provides the set of usual outputs for rating a project: net present value (NPV), benefit-cost ratio and internal rate of return (IRR). It provides also the money value of each impact considered by the (social) cost-benefit analysis, namely of investment cost, operating and maintenance costs, vehicle operating cost, time, safety, user charges, regional and global air pollution. As far as possible, the impacts are disaggregated by impact group, i.e. costs and benefits are assigned for each party concerned by the project, namely, users, operators, developers, the national governments, the European Union and residents. This allows the identification of winners and losers. EUNET allows for sensitivity analyses with respect to the discount rate. Different scenarios affecting the data can be introduced. There is also some possibility to choose between the use of European and national values.

Another module is strictly financial and captures information relevant for a potential private investor. This data is only given for those elements directly under the investor's ownership or franchise: tariffs, transport flows, financial operating and maintenance costs, financial investment costs, and commercial vehicle operating costs. This module provides the private internal rate of return.

From these outputs, the multicriteria analysis module takes the private IRR from the financial module to represent the level of private financing attractiveness criterion. It may also introduce the net present value computed by the (social) cost-benefit analysis A number of non-monetary criteria are included in this analysis, like

- noise, landscape, land take, land amenity, special sites, severance;
- water pollution and local air pollution;
- output, employment and other policy synergy;
- regional accessibility and social cohesion.

These inputs are introduced and scored by the users following suggested definitions. The last two indicators are obtained through a GIS network analysis.

This approach suggests a different point of view with regard to the measurement of the general socio-economic value of a project as opposed to its economic efficiency performance. The EUNET decision-support system provides a wealth of information but no simple recommendation as in a single mono-criterion valuation approach. The decision as to the prioritisation or financing of the projects is left to the users of the decision framework, that is the decision-makers.

One step further in the direction of a decision framework analysis was taken by the CODE-TEN research programme, the objective of which was to 'apply the scenario approach to the study of TEN developments (for freight transports) and the extensions to the CEEC/CIS countries, paying particular attention to the marginal long-term effects and, in particular, the spatial distribution of environmental and socio-economic impacts' (CODE-TEN Technical Annex, 1998). The main output of CODE-TEN has been the development of a comprehensive policy assessment methodology and accompanying decision tools. Given its very wide range of objectives aiming at strategic policy assessment rather than the valuation of specific projects, CODE-TEN naturally aimed at providing a framework for reflection, discussion, assessment and planning towards the formulation of transport policies as part of the general socio-economic policies of the European Union. Important tasks for CODE-TEN were the charting of the infrastructure strategies within corridors as developed by the concerned countries and the formulation of socio-economic scenarios and policy options for the future. These elements together were the object of an assessment followed by policy-relevant recommendations.

Four socio-economic scenarios were formulated combining two hypotheses on two factors: rate of economic growth and the speed of European integration. These hypotheses translated into different origin-destination matrixes for the 500 regions covered by the study. Five future network constellations were examined: the completion of all road projects, the priority given to either rail or road projects along corridors, a complete network for all modes, and a constellation under which no project was implemented. The combination of the socio-economic scenarios and infrastructure developments were assessed with respect to four transport policy scenarios corresponding to different degrees of liberalisation, regulation and the promotion of intermodality.

CODE-TEN relies on the Policy Assessment Model (PAM), an output of the TENASSESS research project, for examining the extent to which a project is in line with the policy objectives from the point of view of the policy owner. It is a test of perceptions not of objective facts. Ten policy aspects are considered:

- the project's congruence with environmental legislation;
- the extent of self-financing of the project based on an estimate of a commercial IRR;
- the degree of the project's intermodality based on the percentage of the flows transported by road;
- the project's contribution to demographic accessibility and regional development based on accessibility indicators;
- the project's importance with regard to international connections as measured by the expected increase in cross-border traffic;
- the project's avoidance of burden for local road traffic and safety aspects;
- the extent to which the project contributes to the more general goals of liberalisation and privatisation;
- time saving in relation to capital cost.

These indicators are scored on a scale from - 5 to + 5 for each combination of scenarios and infrastructure strategies, and they are aggregated by weighted summation. The weights vary with the relevant transport policy scenarios; another set of weights purports to assess the acceptability of a transport initiative by the concerned countries. The use of a different set of weights permits a sensitivity analysis of the results. The acceptability test is applied separately for each country involved and presented on geographical maps. Altogether, the CODE-TEN strategic assessment methodology represents a decision support system for transport policies. It is emphasised that 'individual final scores are not, in themselves, the main output (...) Rather, the policy assessment model is an analytical tool, designed to test a variety of options, refinements and weights in order to generate a measure of the overall worthiness and robustness of a transport initiative' (Bulman and Brown in Giorgi and Tandon, eds, CODE-TEN Final Report, 2000).

CODE-TEN also examines whether there are barriers to the implementation of transport projects. This analysis uses the TENASSESS Barrier Model which is based on the decision-tree approach. It involves the charting of the projects on any particular corridor according to their phase of implementation and the number and type of obstacles they are facing. The combination of these results with those of the above analysis provides an indication of the likelihood that an investment programme will be implemented as is or whether it will need some prior re-design.

Subsequently freight traffic assignments are carried out for the different combinations of scenarios and infrastructure developments. This assignment is made with the help of a GIS analysis based on an extensive data bank which includes information on the origins and destinations of flows per type of commodity, the capacity of links, and transport costs. The assignments are made successively for the different types of commodities according to their value, the cost of transport being adjusted at each step as a function of the congestion on the links. Traffic impacts in each corridor can be computed from these results. Finally, CODE-TEN proposes a strategic assessment of the corridor development alternatives using the EUNET multicriteria methodology.

It is worth noting that no (social) cost-benefit analysis is proposed and that uncertainty is only dealt with through sensitivity analysis using scenarios. Given the strategic planning and exploratory outlook of an analysis at the level of corridors, these two abstentions are understandable. Cost-benefit analyses of each corridor configuration would presumably require additional information not available at this time, despite the remarkable efforts made by both EUNET and CODE-TEN in gathering data. Further efforts in that direction would be worthwhile, particularly for the continued development of a detailed Trans-European network model that would allow deeper economic analyses of corridors and other initiatives and particularly of their inter-regional network effects. Likewise deserving further attention is the uncertainty characterising impacts of corridor programmes at the local and regional level.

CODE-TEN provides a rather comprehensive analysis framework for planning transport initiatives. It views the infrastructure programmes from different points of views: their degree of congruence with policy objectives, their likelihood of implementation considering administrative and legal obstacles as well as concerned population interests, inter-regional traffic impacts, and a global multicriteria analysis of social, environmental and economic impacts. It does not aim at giving a simple recommendation to a 'fictitious' decision-maker, but rather a tool for reflection and discussion by planners and the numerous political decision-makers.

Multicriteria Analysis and Uncertainty

The Need to Analyse Projects' Uncertainty

The outcome of a project assessment – its net present value for instance as estimated by a (social) cost-benefit analysis or an equivalent value provided by a multicriteria analysis – is often affected by some degree of uncertainty. Three main reasons explain the presence of risk or uncertainty in decision making: the inevitable imperfection of statistical estimations, the difficulty of estimating and forecasting within incomplete and/or dynamic systems, and the basic ignorance of what the future holds (recession, war, accident, etc.). Many of the problems reviewed in the first part of this paper as well as the practical solutions proposed by advanced (social) cost-benefit and multicriteria analyses are obvious illustrations of how uncertain the forecast impacts can be. If the decision-maker is risk averse, he or she must identify and seek to account for all these risks in the process of decision making. Particularly relevant in this connection is the size of the project compared to the country's gross income.

For small projects that are not correlated with the country's gross income, Arrow and Lind (1970) have shown that the risk could be neglected if it is actually shared by a very large number of citizens. In this case, the state, which plays a role similar to a mutual insurance fund, can make decisions without attention to risk, i.e. it can take a risk-neutral attitude in decision-making. However, this neglects the fact that, besides the usual financial costs and benefits, there may be substantial external effects that are not as equivalently distributed over the total population, or not distributed at all but affect everyone in the same way (Fisher, 1974). These are also components of the global risk of a project that should be taken into account. Furthermore, there will always remain a 'systematic risk' which pertains to the possibility of an economic recession, a tornado or a war. As far as such risks are concerned, most project outcomes are correlated to these events so that this part of the global risk cannot be spread over the population.

On the other hand, the risk of a large-scale project cannot be neglected, particularly if its impacts are not evenly distributed over the population, be it with regard to socio-economic groups or regions. The equity aspect of large-scale projects determines its level of risk as well as who bears this. Additional risk is generated through the interdependence relations between this and other projects as well as the effect the project has on the entire economy.[3] The outcome of a large project is also affected by the systematic risk.

Methods to Analyse Risk

For analysing risk in decision making, the expected utility theory of von Neumann and Morgenstern (1944) is still the reference model. It assumes that the utility function is strictly concave if there is any degree of risk aversion. It follows that the expected utility, taken as a measure of a risky project's utility, has a lower value than the utility of a certain project with outcome equal to the weighted average outcome of the project. Hence, there is a loss of utility resulting from the project's risk. Furthermore, if there is a relation between utility and income, it is possible to compute equivalent money values and the cost of the risk involved in a situation or a project. This cost amounts to the difference between the money value of the average outcome of the project and that of the certain project that has a utility equal to the expected utility of the uncertain project. In the financial literature, this difference is called the Risk Premium since it is the maximum that one would pay for insurance against the risk involved.[4] It is a measure of the willingness to pay for averting that risk.

A linear utility function may well be the most convenient form for a multicriteria analysis but it implies risk neutrality. In order to avoid this strong assumption, a non-linear form must be used. Keeney and Raiffa

(1976) have proposed a methodology based on an additive non-linear specification. It is important to note that the specification of an additive utility function is based on the assumption of mutual utility independence. This means that if two projects are characterised by the same values for the same criteria, the preferences between them do not depend on these given values but only on the level of the remaining criteria. This is a rather strong condition which suggests, for example, that the willingness to pay for reducing pollution does not depend on the level of regional development. This is indeed generally the case, I would contend. However, since we are concerned only with the additional utility obtained from implementing a particular project, it is easier to meet this condition of mutual independence than it may seem at first. For instance, we may think that the additional utility obtained from a project reducing pollution does not depend as much on the project's impact on regional development as it does on the level of regional development. Also, it can be argued that the utility provided by a project can be seen as a utility differential, which is defined mathematically as a weighted addition of marginal utilities, i.e. impacts' utilities.[5]

Let us note that there are basically two types of utility functions. The usual one, which is applied in most texts of micro-economic theory, is derived under conditions of certainty. It expresses the preferences of the decision-maker when confronted with certain outcomes. In the more advanced theoretical literature on uncertainty, this is called the 'value function'. When a value function is used for computing an expected utility, it implies that the utility of the uncertain outcome is the weighted average of the utilities that would be obtained from certain outcomes (Bernouilli, 1738). From an operational point of view this means, concretely, that to estimate such a value function we may focus only on the decision-maker's preferences between projects with certain outcomes. In contrast, a 'utility function' incorporates whatever attitude towards risk taking the decision-maker may have in mind: risk aversion, risk neutrality or risk proneness. Actually, it can be thought of as a transformation of the value function that accounts for the attitude towards risk in the assessment of the alternatives confronted by the decision-maker. It implies that, when estimating the utility function, questions addressed by the analyst to the decision-maker must refer to uncertain outcomes, or lotteries, as uncertain outcomes are more generally referred to in the literature. This is obviously a much more difficult task, particularly in an analysis involving several criteria.

A possible procedure would be as follows: first, build a value function of some form; then transform this into a utility function by questioning the decision-maker on the utility he or she attaches to different lotteries of values. Naturally, this procedure neglects the possibility that the decision-maker's risk attitude varies across criteria. In any case, given the current practice of public project valuation, we think that an important step forward would already be achieved if an attempt were made to obtain at least non-linear value functions. While the methodological distinction between a value and a utility function is assumed in order to avoid theoretical confusion and controversy, it will not be necessary to maintain such a distinction throughout our discussion. In what follows, then, only the more common terminology of a utility function will be used.

Two approaches for generating additive multi-attribute utility functions can be distinguished. A number of methods start by separately building partial utility functions, and, subsequently, estimate the weights that link these functions. This approach was initially proposed by Keeney and Raiffa (1976) and can be implemented in different ways. The more recent MACBETH method by Bana e Costa and Vansnick (1994, 1997) is probably one of the most convenient as it applies a questionnaire methodology with verbal propositions that seeks a good approximation of interval-scaled preferences in certainty. Such a value function measuring strength of preference cannot, however, be taken as a real utility function for valuing risky actions, since asking questions in certainty cannot extract risk attitudes.[6]

Other methods proceed from stated global preferences between projects with linear goal programming models. Among these, the UTA method proposed by Jacquet-Lagrèze and Siskos (1978, 1982) is probably the most developed and useful. Its general framework can also be used to build separately and assemble non-linear partial utility functions (Beuthe *et al.*, 1998). A software, called MUSTARD (Scannella and Beuthe, 1999), has been developed which proposes several variants of the UTA model under both conditions of certainty and uncertainty, with a global assessing of the multicriteria function as well as with a separate estimation of the partial utility functions. It includes a questionnaire of preferences between projects and lotteries of projects which allow the estimation of attitudes towards risk. In the end, it computes equivalent money values and risk premiums in cases where there is a money value criterion.

Besides this rather classical economic theory approach to the problem of uncertainty, one should consider yet another more recent approach based on the use of fuzzy numbers. Under this approach, the impacts are taken to be uncertain over a range of values, but no stochastic distribution is postulated. While the classical approach supposes that the decision-maker has a clear system of preference embodied into a utility function, the fuzzy numbers model of decision making does not. On the basis of a system of given weights, 'fuzzy' preferences are aggregated over all criteria to produce a global index which indicates the strength of preference for one of two projects. These can be combined to rank all projects according to a complete or partial pre-order. There are a number of such decision models proposed in the scientific literature (cf. Williams *et. al.*, 1998, and Beuthe *et al.* 1998).

The fuzzy numbers approach has the advantage that it does not require the decision-maker to always have clear-cut preferences, and does not need an estimation of probability distributions. On the other hand, it requires the setting of a certain number of (arbitrary) parameters to define the fuzzy preferences system; and it does not permit to compute equivalent money values and risk premiums, since it does not provide any precise rate of substitution between criteria and money. Nevertheless, it is an excellent tool to investigate the decision-maker's preferences between projects and transport initiatives.

All these approaches naturally meet the problem of the 'fictitious decision-maker' and of the multi-level decision context, particularly at the European level. However, uncertainty is so pervasive in project assessment that the possibility of proceeding any further into risk analysis must be seriously considered. To begin with, whatever the selected methodology, it is necessary to proceed to a thorough analysis of all the factors which affect the outcome of a project. We mentioned earlier that a sensitivity analysis is unable to solve entirely the problem of uncertainty since it leaves the decision-maker with the problem of deciding how much risk is worth taking. A sensitivity analysis performed without a preliminary research of the possible future scenarios and of the possible range of variation of the different impacts and of their associated probabilities would remain substantially incomplete. Whenever possible, the likelihood of each hypothesis, scenario or valuation should be assessed and submitted to decision makers.

This task is not an easy one. Some sophisticated methods have been proposed to generate the probability distribution of a utility function, for instance by a Monte-Carlo approach.[7] They are probably too difficult to apply in many cases of project assessment. More practical 'down-to-earth' methods are used by risk managers in the insurance field. Some are based on ex-post assessments and they suggest that ex-post analyses of project assessments would produce valuable information for improving future assessments with respect to biases and unexpected outcomes. A policy of systematic ex-post analyses is certainly to be recommended.

Some other methods are more tailored to ex-ante appraisals. Among these, the so-called 'Activity Centres' approach proposes the division and distribution of the problem of risk to sub-groups of analysts who can study a specific risk in more depth. Each analyst involved in the valuation of a project is questioned about his or her estimation, or perception, of the possible realisations of the variables and/or criteria they are experts about: the likely range of the variables, the type of distribution and its mean or mode, and whatever information is available about the probabilities. A simple questionnaire can be set up by the research team to facilitate this approach: its design would induce a global reflection on the risks involved in a project and in its assessment, and its application to each specific factor would help the concerned analyst to provide appropriate answers. Examples of this approach were given recently by Brod (1993), Lewis *et al*. (1995), and by Beuthe *et al*. (1998, 2000).

For the sake of completeness, two well-known approaches to handle the problem of uncertainty must still be mentioned. The first one is the use of a 'risky' discount rate, which is a computationally convenient method. However, its definition is open to many questions. Furthermore, it increases the discount rate, hence decreases more than proportionally the weight of later years. Its use implies that the risk increases with the age of a project. Not only does it reduce the weight of the positive effects, but also the weight of the negative effects. Altogether, it does not appear as a satisfactory procedure.

The second approach is based on the use of the 'pay-back period', i.e. the period of time which is necessary to recoup the full cost of the project. A longer pay-back period is then seen as a factor of higher risk. But this approach neglects the risk during the pay-back period and the outcomes of the project afterwards. Hence, it creates a bias against long-duration projects. This is only an indicator of the short-run rate of return rather than an indicator of risk. In any case, it may provide a useful information to the decision-maker, even though it does not help much in deciding whether the risk involved in a project is worth taking.

Equivalent Money Values and Risk Premium in Multicriteria Analysis

As deriving equivalent money values in a multicriteria analysis context is not usually done, an appropriate methodology still remains to be set. The following paragraphs outlines a methodology that is founded on sound theoretical considerations.

Let $U(A) = u(M) + v(X)$, the additive utility function of a project A defined by two variables: M, its net present value, and X, its environmental impact measured in physical units. This function is illustrated in two dimensions in the following diagram where the (indifference) curve is defined by a given level of utility. The equivalent money value (EMV) of the project which delivers this utility corresponds to M(A), the intercept of indifference curve $U°(A)$ on the M axis.[8]

Figure 7.1   Equivalent Money Value

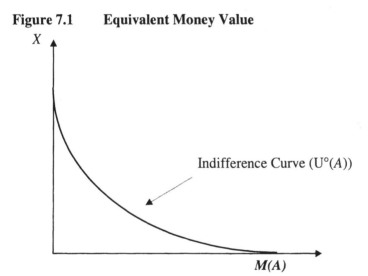

In a situation of uncertainty, with random outcomes, the equivalent value is also random. However, a certain equivalent money value M* can be computed, which is defined by the condition that its utility is equal to the expected utility of the random M(A). Indeed, the decision maker would then be indifferent between the amount M* and the project with random outcomes. The risk premium (RP), which corresponds to the cost of uncertainty, is then equal to the difference between the EMV computed at the average values of M and X, and the certain M* of the project, i.e. RP = M(Ā) - M*. The risk premium can be interpreted as the maximum amount that the decision-maker would be ready to pay in order to escape the uncertain situation.

In practice, rather than computing these values by simple extrapolation of the utility function, it is proposed to transform utilities on the basis of its average slope over its variation interval (Beuthe *et al.*, 2000).

With equivalent money values, it is then possible to keep unchanged the basic rules for accepting a project in (social) cost-benefit analysis: a project can be accepted if its net present value – all impacts' values included – is positive. If choices must be made between a number of projects, the ratio of the net present value to the budgetary cost of each project should be used as a rank-ordering index.[9] Its use will help maximise the total net present value of all projects which can be implemented with the available total budget.

Regardless of the method used to account for the risks inherent in a project, a proper analysis requires systematic estimation of the uncertainty affecting all relevant variables and parameters including their range of variation, and the probability of their realisation. If statistical estimates are not available, these elements should be assessed by the analyst, in some cases with the help of outside experts, on a subjective basis. Since the multicriteria function is taken to be additive, each distribution is assumed to be independent. Nevertheless, account should be taken, as much as possible, of all factors that affect their distribution and, particularly, of the so-called systematic risk. The latter, which corresponds, for instance, to the possibility of an earthquake or of an economic recession, is as much a part of the overall risk as any other factor specific to a project.

Considering the difficulty of the task, it is necessary to limit the requested information to a minimum. For that reason, Beuthe *et al.* (2000) have suggested the use of triangular distributions which can approximate normal distributions but can handle also cases where the distribution is not symmetrical. This can be a useful approximation of many continuous distributions with the rectangular one as a limit case. The option of a discrete distribution should also be open as there may be qualitative criteria with but a few values.

Socio-Economic and Environmental Criteria

Development Impacts and Accessibility

There are different types of accessibility indicators. The simplest ones, which can be called topological measures, describe the transport network existing in a zone; they are only measures of accessibility between different points of the same zone. The bilateral indicators relate to two different zones. They may take different forms like a (weighted) sum of the populations which can be reached within a given travelling time or a gravity indicator, often a product of the populations or economic activities in two zones weighted by an attraction-decay function based on the money or time cost of transport between these two zones. These bilateral indicators may then be aggregated with respect to all destinations or zones involved in the analysis or a subset of them, for instance the destinations which can be reached within two hours or the destinations the population of which correspond to a given market.[10] There are many variations around these basic formulas with more or less sophisticated weighting systems. An interesting example is the POINTER formula (Kronbak, 1998) which is based on the product of the population at the destination and the time remaining for interaction with the population, for instance eight hours of working time minus a four hours total travelling time. Differences between indicator values with and without a project are used to model an accessibility impact resulting from a project.

Aggregated indicators can be used for describing a region's new transport interaction opportunities arising from a project implementation. The meaning of the indicator depends to some extent on the destinations which are included in the aggregation. For instance, the relevant travelling time limit for business travellers is different from the transport time limit for freight, and the latter may vary with the type of commodity or service. Hence, whenever an accessibility indicator is introduced in a multicriteria analysis it is important that the decision-maker be made aware of its definition.

Obviously, the impacts measured on the base of these indicators may be highly correlated with the corresponding total value of time and cost savings which are included in the (social) cost-benefit analysis. Actually one may fear a double counting of the same effect if a simple global accessibility impact for the whole of Europe is introduced next to the monetary result of a (social) cost-benefit analysis. In particular, one should keep in mind that the value of time saving in cost-benefit analysis is computed on the basis of a shadow price estimation which in principle takes into account the circumstances of transport. There would in fact be a substantial double counting, and this is the reason why such an impact is not introduced separately in the Dutch and German appraisal framework, nor is it included in the EUNET (social) cost-benefit analysis.

However, accessibility indicators may have some merits when they are computed separately for each zone considered in the study, as was done for the analysis of the Øresund bridge (PRODEC, 2000). In the latter case they describe the new transport opportunities offered to the different regions, particularly when they are presented with easy-to-read maps. Their integration in a multicriteria analysis remains problematic though, unless they are somehow aggregated according to some regional weighting scheme which would reflect the regional policy objective of the decision-maker. In EUNET, two such accessibility indicators will be introduced in the multicriteria analysis module: the first one, with weights varying inversely with the state of economic development, would be used to compute the projects impact on social cohesion under the assumption that improved accessibility induces economic development; the second, with weights inversely related to Euclidean distance, would be used to compute an impact on spatial accessibility. A possible alternative to the first indicator, but only if the objective is equal accessibility for all regions, is to use a Gini index which is a measure of inequality.

Nevertheless, there is no escape from the fact that accessibility indicators can have several meanings so that their indications often remain ambiguous. For instance, as a result of the Øresund bridge, the POINTER indicator shows an improvement of accessibility which is higher for less populated areas, in this case Southern Sweden as opposed to the Copenhagen conurbation. Obviously, people and enterprises in Southern Sweden will obtain relatively higher transport and travelling opportunities since they will be able to reach more easily the densely populated Copenhagen area. On the other hand, the attraction of that area will certainly increase and that may induce a negative longer-run effect on some of the local activities in Southern Sweden but also in more Northern parts of Sweden. Actually, it should be recognised that relative changes in accessibility may result in the redistribution of activities rather than in the creation of new activities which would correspond to real growth.

Thus, if an accessibility criterion is introduced at all in a multicriteria analysis, it is important to define with precision what it is meant to represent or to which policy objective it corresponds. Otherwise, it is not possible to attribute a sensible weight to an accessibility criterion. If accessibility is taken to be about convenience or cost saving the corresponding regional impacts could be measured by an index like POINTER. If, on the other hand, accessibility is taken to reflect the concern about equity, a Gini index of unequal regional economic development would be more appropriate. As an instrument of political integration, some kind of a global accessibility indicator could be used, but it should be made explicit that the objective is political and not economic. As an instrument of economic integration (common market), the inclusion (or exclusion) of accessibility as a separate criterion should depend on the modelling of future transport flows in order not to face the problem of double counting.

Accessibility is intuitively an appealing concept, but the economic and sociological effects of regional accessibility are many and not yet very well understood. The role of accessibility as a means to promote regional economic activities by an improved network remains doubtful. There are still few satisfactory empirical investigations of this question and they provide uncertain (and controversial) results. Indeed a recent review of researches on this topic by Banister and Berechman (2000) concludes that

> in developed countries, where there is already a well-connected transportation infrastructure network of a high quality, further investment in that infrastructure will not on its own result in economic growth. Transport infrastructure acts as a complement to other more important underlying conditions, which must be also met if further economic development is to take place (p.318).

These conditions are a favourable economic environment, availability of funds, supporting legal, organisational and institutional policies and processes, plus complementary policy actions. Actually, a package of development measures is needed in order to obtain effective development. From that point of view, an analysis focusing only on transport programmes is much too narrow. It is somewhat analogous to an infrastructure assessment which does not take into account part of the investment cost.

Similar comments can be made about some other criteria introduced to indicate development impacts, like 'increment in GDP' and 'increment in employment' both included in the EUNET multicriteria analysis. From a redistribution point of view it is advisable to take into account which regions benefit from an employment increment. This comment raises the question of which effect is really aimed at: general development or better regional distribution of development? Moreover, a rather macroeconomic computation of employment and output increments, as sometimes proposed in national practices, raises the question whether better results could not be obtained with the same global expenditure by investments outside the field of transportation. Actually, introducing such criteria tends to enlarge the field of investigation to other projects which could be envisaged by the public authorities.

With regard to the problem of double counting which is raised forcefully by the consideration of such criteria, one should perhaps first ask whether it is at all possible to disentangle the effects covered by the criteria of 'output', 'employment', 'spatial accessibility' and 'social cohesion'. One must also wonder whether research teams are not perhaps a bit too much willing to specify a criterion for each of the objectives specified in European policy documents, even when these are very general and vaguely defined. A more honest and perhaps also better approach would be to admit that background available knowledge is lacking or that it is neither theoretically nor empirically possible to observe and measure all impacts separately.

We can conclude that a better understanding of the role played by transport infrastructure and accessibility in regional development is called for. In the meantime the use of accessibility indexes and other such criteria in project assessment should be carried out with caution, despite the fact that these are convenient and rather popular concepts among policy designers.

Environmental Impacts

The field of environmental impact assessment has been very active for some time. In several countries and at the European level – with programmes like ExternE, COPERT II, MEET and COMMUTE – attempts were made to value external effects produced by different pollutants in money terms and, subsequently, to incorporate these into (social) cost-benefit analyses. It is beyond the scope of this paper to review these methodologies at length and in depth. This would require a full lengthy paper of its own, and the present one is more concerned with the general 'philosophy' of assessment.[11]

The reader may guess that the present author totally agrees with these attempts to attach money values to such effects. Nevertheless the problem of uncertainty affecting such valuations needs to be more seriously addressed and as such deserves mentioning here. This is well illustrated by the following table which gathers the estimates of pollutants emissions found in various studies. Naturally, they may concern different conditions of operations and environments and refer to different types of vehicles. Still they show a very wide range of variation. These are estimates of physical emissions which must still be transformed in money values, which should still further increase the range of variation. The same large variation is observed by an OECD report (1997).

Figure 7.2   Pollutant Emissions (grams per tonne-km)

	Source	PM	NO_x	VOC	SO_2	CO	CO_2
Road							
	TRENEN	0,033	0,390	0,016	0,078	0,190	82,554
	Hickman (light truck)	0,07726	0,82207	0,21575	-	0,36175	89,0477
	Hickman (heavy truck)	0,03064	0,60956	0,06176	-	0,13354	45,9559
	Forkenbrock	0,103	1,371	0,086	0,039	-	-
	INA	0,13	1,7	0,1	0,045	0,3	83
Rail							
	TRENEN	0,015	0,179	0,007	0,032	0,075	33,945
	RIVM (electric)	0,00047	0,02	0,00047	0,0077	0,002	24,44
	COST319 (diesel)	0,02	0,39	0,05	0,05	0,08	9
	INA	0,002	0,13	0,003	0,007	0,012	7,5
Water							
	TRENEN	0,020	0,232	0,009	0,041	0,098	44
	PLANCO	-	1,2	0,08	0,01	0,2	63,5

Source: Table prepared by F. Degransart (2000)

Global air pollution, for example, is valued in EUNET at 50 ECU per tonne, whereas in ExternE at 41 ECU per tonne. The ExternE estimation was subsequently revised down at the range of 1-30 ECU per tonne. As noted by Nellthorp et al. (1999), 'The uncertainty associated with all these values, arising from assumptions made in their derivation (e.g. the choice of discount rate, or the type of ecosystem model) could be as great as +/- 95%' (Vol. 1, p. 21).

The following table prepared by Hammische (1997) gathers the final estimates of different external effects, which can be found in a set of (mainly) French studies. They also exhibit a wide range of variation.

Figure 7.3   External Costs of Road Transport of Freight

Source	Bruit	Pollution	Accident	Congestion
Josse (1989)			6,1 ctsF85/veh.km	0,591 FF85/veh.km
Brossier (1991)	0,02 à 0,06 FF/veh.km	0,1à1,35 FF/veh.km	0,1377 FF90/veh.km	0,413 FF90/veh.km with dense traffic 1,261 FF90/veh.km on congested highway
Quinet (1991)			7,7 ctsF85/veh.km	
LET (1995)	0,027 à 0,021 FF/veh.km	0,21 à 0,843 FF/veh.km	0,052 à 0,096 FF/veh.km	0,256 à 0,344 FF/veh.km
Boiteux (1994)		5,6 ctsF92/tkm utile		0,125 hour/veh.km in congestion
INRETS (1996)	0,4 à 0,6 ctsF91/veh.km en rase campagne	3,1 à 7,3 ctsF91/veh.km dont 0,7 à 2,5 pour l'effet de serre	2,7 ctsF91/veh.km	
CEE (1987)	0,6 ECU/1000 t.km	7,8 ECU/1000 t.km	8,2 ECU/1000 t.km	8,1 ECU/1000 t.km
OCDE (1990)			6,7 à 9,8 ctsF/km	
Grupp (1986)		0,18 à 0,82 ECU/100 t.km		
Ecoplan (1989)		4,66 ECU/100 t.km	0,028 ECU/veh.km	
Prognos (1993) PLANCO		2,15 ECU/100 t.km 1,33 ECU/100 t.km	0,0092 ECU/veh.km	

These tables show that, despite considerable efforts over quite a few years of research, an important uncertainty characterises all these estimates. The question is, therefore, whether sensitivity analysis is sufficient to deal with this aspect of projects' assessment. In any case, if for all kinds of understandable practical reasons, this is the only method which can be applied, such an analysis should be applied to many more valuation parameters than is usually done.

Conclusions

In the course of reviewing the recent evolution of assessment methods applied to transport infrastructure projects, we showed that in its quest for a comprehensive analysis of all impacts of projects, (social) cost-benefit analysis has come to accept that some valuations in monetary terms be based on political judgement rather than on market values. We argued then that, since multicriteria analysis was an appropriate methodology to assess in relative terms the worth for the decision-maker of some more qualitative or political criteria, there was no reason not to use these relative weights to transform utilities into money equivalent values. This appears to be particularly appropriate when there is a budget constraint, what necessarily raises the question whether a project's outcome is worth the spending of a set budget. An advantage of enlarged money valuation is that it allows to handle the selection and the ranking of projects on the basis of the usual rules of cost-benefit analysis. More generally, to be able to state that such a qualitative effect has an equivalent money value amounting to a certain sum and to compare this sum with other monetary values should certainly introduce a useful and concrete reference facilitating the decision-maker's appraisal effort. For most people money values are much easier to understand than utilities. I proposed a theoretically sound methodology for transforming utilities into equivalent money values in the context of multicriteria analysis.

Another theme of concern was the treatment of uncertainty. The current practice is often limited to a rather casual sensitivity analysis applied on a few parameters, like the discount rate in (social) cost-benefit analysis, or the weights of the multicriteria function. The need for a thorough analysis of the risks involved in a project and of the parameters uncertainty was underlined. Various methods were mentioned for handling uncertainty. I would favour a methodology based on the expected utility theory, which permits the computation of a project's risk premium, if transforming utilities in equivalent money values is deemed acceptable. In any case, regardless of the method adopted, the first step to be taken should be a systematic analysis of all the specific risk factors. That alone would allow a real appraisal of the risk taken by the decision-maker when recommending the acceptance of a project.

In the last part of the paper, several wider ranging decision criteria were reviewed, like regional development, employment and accessibility. The enticing inclusion of these economic or political criteria raises a number of problems concerning their rigorous definition, the danger of double counting, and the reality of the link between the infrastructure projects and their purported effects. One may fear that there is a certain amount of unsubstantiated wishful thinking involved in their inclusion. More research is certainly needed in this area.

Notes

1 I am pleased to acknowledge the useful comments received from an anonymous referee.
2 This is, for instance, the problem in the EUNET MCA framework. There, pair-wise comparisons in 'Analytical Hierarchy Process (AHP)' fashion (Saaty, 1980) are used to build up a system of criteria weights. For doing so, the user is expressing his binary preferences on a - 8 to + 8 scale. In contrast with AHP, this process is not linked to the scoring of the projects' impacts by similar binary comparisons.
3 A detailed discussion of these issues can be found in Chapters 6 and 7 of Beuthe and Scannella (1995).
4 There is an analogous concept, the option price, which represents the willingness to pay when the effects of an uncertain project are compared to an uncertain initial situation.
5 For a more detailed discussion, see Beuthe (1994), Chapter 6 of the APAS/Road/3 report.
6 On this topic, see Bell and Raiffa (1988) and Bouyssou and Vansnick (1990) who discuss under which restricted conditions equivalence of both types of functions can be found.
7 For a brief review and references of various methods, see Beuthe *et al.* (1998).
8 Given the specification, indifference curves have an intercept on each axis.
9 Strangely enough this rule for ranking projects and allocating budget does not seem to be well known by practitioners of SCBA. Actually, there is a difference between the budgetary cost Cb which must be allocated between projects, and the resource cost Cr used to compute a project's NPV. Cr may have been computed on the basis of shadow prices without any relation to budgetary cost. Hence the ratio B/Cr, although a useful heuristic, is not a correct reference for prioritising projects for budget allocation. For a complete treatment of the matter involving comparisons of different subsets of projects see Weingartner (1963).
10 Useful references are CESUR-IST (1999) CODE-TEN D5 and Chatelus (1997).

11 An excellent review of the results available in this field for assessment use is given in Nellthorp *et al.* (1999).

References

Banister, D. and Berechman, J. (2000), *Transport Investment and Economic Development*, U.C.L. Press, Taylor and Francis, London.

Bell, D.E. and Raiffa, H. (1988), 'Marginal Value and Intrinsic Risk Aversion', in Bell, D.E., Raiffa, R. and Tversky, A (eds), *Decision Making: Descriptive, Normative and Prescriptive Interactions*, Cambridge University Press, Cambridge, pp.384-397.

Bernouilli, D. (1738), 'Specimen Theoriae Novae de Mensura sortis', Proceedings of the Imperial Academy, St. Petersbourg, translated into English as 'Exposition of a New Theory on the Measurement of Risk', Econometrica, 1954, Vol. 22, pp.23-36.

Beuthe, M. (1994), 'Evaluation', Chapters 6 and 7 of *Final Report of APAS /ROAD /3*, Research programme of European Commission DG VII- A4, Brussels.

Beuthe, M., Eeckhoudt, L. and Scannella, G. (1998), *Uncertainty in Project Assessment*, EUNET Deliverable D10, Vol. 2, Appendix III.

Beuthe, M., Eeckhoudt, L. and Scannella, G. (2000), 'A Practical Multicriteria Methodology for Assessing Risky Public Investments', *Socio-Economic Planning Sciences*, Vol. 34, No. 2, pp.121-140.

Beuthe, M., Grant-Muller, S., Pearman, A. and Tsamboulas, D. (1998), 'Prioritising Trans-European Network Transport Initiatives', Paper presented at the 8th WCTR Conference in July 1998, Antwerp.

Bouyssou, D. and Vansnick, J.C. (1990), 'Utilité Cardinale' dans le certain et choix dans le risque', *Revue Economique*, Vol. 6, pp.979-1000.

Brod, D. (1993), 'Analysis of the Financial Viability of the new Denver Airport, in Infrastructure Planning and Management', *Proceedings of a Conference of the American Society of Civil Engineers*, pp. 432-441.

CESUR – IST (1999), *Assessment of Spatial and Socio-Economic Impacts*, CODE-TEN Deliverable, CESUR, Lisbon.

Chatelus, G. (1997), *Accessibilité Interrégionale,Théorie et example d'application à l'échelle Européenne*, Rapport INRETS No. 217.

Cost 319 (1999), *Estimation of Pollutant Emission from Transport*, Final Report, European Commission, Directorate General Transport, Brussels.

Degransart, F. (2000), *Rapport de synthèse sur les externalités de transport*, FUCAM, Mons.

Forkenbrock, D.J. (1999), 'External Costs of Intercity Truck Freight Transportation', *Transportation Research A*, Vol. 33, No. 7/8, pp.505-526.

Giorgi, L. and Tandon, A. (eds) (2000), *The Decode Method* (Final report), CODE-TEN Deliverable 7, ICCR, Vienna.

Grant-Muller, S., Nellthorp, J., Chen, H., Pearman, A. and Tsamboulas, D. (1999), *Decision Analysis Report and Prototype*, EUNET Deliverable D16, EUNET Consortium, Leeds and Brussels.

Halcrow (1998), *Assessment Methodology Report*, TENASSESS Deliverable, D4, Halcrow, London.

Halcrow (2000), 'Application of the Policy Assessment Model to the Strategic Evaluation of Development Alternatives', CODE-TEN, internal working paper.

Hämäläinen, R.P. and Pöyhönen, M. (1996), 'On-Line Group Decision Support by Preference Programming in Traffic Planning', *Group Decision and Negotiation*, Vol. 5, Nos. 4-6, pp.185-200.

Hickman, A.J. (1997), 'Emission Function for Heavy Duty Vehicles', *TRL Report*, No. PR/SE/289/97, Cowthorne.

ICCR (2000), *CODE-TEN Scientific Report*, Deliverable R9, ICCR, Vienna.

International Navigation Association (INA), *Bateaux de navigation intérieure et pollution*, Rapport du groupe de travail 14.

Keeney, R.L. and Raiffa, H. (1976), *Decisions with Multiple Objectives: Preferences and Value Tradeoffs*, Wiley, New York.

Layard, R. and Glaister, S. (eds) (1994), *Cost-Benefit Analysis*, Cambridge University Press, Cambridge.

Leleur, S. (2000), *Road Infrastructure Planning*, Polyteknisk Forlag, Lyngby.

Lewis, D. (1995), 'The Future of Forecasting: Risk Analysis as a Philosophy of Transportation Planning', *TR News*, 177, pp.3-9.

Mackie, P., Palmer, A., Pearman, A., Watson, S. and Whelan, G. (1994), *Cost-Benefit and Multicriteria Analysis for New Road Construction*, EURET 1.1 Concerted Action Report.

Malczewski, J. (1999), 'Spatial Multicriteria Decision Analysis', in Thill, J-Cl. (ed.), *Spatial Multicriteria Decision Making and Analysis*, Ashgate, Aldershot.

Nellthorp, J., Mackie, P., Bristow, A. (1998), *Measurement and Valuation of the Impacts of Transport Initiatives*, EUNET Deliverable D9, Vols. 1 and 2.

OECD (1997), *Marchandises et environnement: les effets externes de la libéralisation des échanges et des réformes dans le secteur des transports*, GD(97)213, OECD, Paris.

Pearce, D.W. and Nash, C.A. (1981), *The Social Appraisal of Projects: A Text in Cost-Benefit Analysis*, Macmillan, London.

PLANCO (1994), *Environmental Impacts of Transport Infrastructure*, PLANCO, Essen.

PLANCO (1999), *Transport Information Management System*, CODE-TEN, Deliverable D2, PLANCO, Essen.

PRODEC (2000), *The Strategic Impacts of the Oresund Bridge*, Oresund Konsortiet, Copenhagen.

Rijksinstituut voor Volksgezondheid en Milieuhygiëne (1998), *Verkeer and Vervoer in de Nationale Milieuverkenning 4 1997-2020*, Bilthoven.

Saaty, T.L. (1980), *The Analytical Hierarchy Process*, Mcgraw-Hill, New York.

Scannella, G. and Beuthe, M. (1999), *MUSTARD User's Guidebook*, FUCAM, Mons.

Schmid, A.A. (1989), *Benefit-Cost Analysis*, Westview Press, Boulder.

Shakun, M.F. (1996), *Modelling and Supporting Task-Oriented Group Processes: Purposeful Complex Adaptive Systems and Evolutionary Systems Design*, Group Decision Making, Vol. 5, Nos. 4-6, 5-17, 1996.

Stiglitz, J.E. (1988), *Economics of the Public Sector*, W.W. Norton.

Thill, J-Cl. (ed.) (1999), *Spatial Multicriteria Decision Making and Analysis*, Ashgate, Aldershot.

TRENEN (1995), *Final Report*, Joule Program of the European Commission, New York.

Tsamboulas, D. and Mikroudis, G. (2000), 'EFFECT – Evaluation Framework of Environmental Impacts and Costs of Transport Initiatives', *Transportation Research*, Vol. 5 D, No. 4, pp.283-303.

Weingartner, H.M. (1963), *Mathematical Programming and the Analysis of Capital Budgeting Problems*, Prentice-Hall, London.

Williams, I.N., Mackie, P.J. and Tsamboulas, D. (1999), 'Assessing the Socio-Economic and Spatial Impacts of Transport Initiatives: The EUNET Project', *Proceedings of the 8th WCTR Conference in Antwerp*, Pergamon, Vol. 4, pp.99-112, Antwerp.

8 Criteria for Evaluation Towards Sustainability

KLAUS RENNINGS AND SIGURD WEINREICH

Introduction

This book deals with the context of project and policy evaluation and how this influences the definition of the problem, the selection of evaluation methods and their application. Against this background, the specific objectives of this chapter are to answer the following questions:

- What are sustainability criteria?
- Which sustainability indicators are available and how can sustainability criteria be derived from different indicator concepts?
- Which evaluation methods (indicators, cost-benefit analysis, cost-effectiveness analysis, multi-criteria analysis better fit sustainability criteria?)
- How should evaluation methods be modified in view of sustainability criteria?
- Can sustainability criteria be formulated in an operational way for policy applications (e.g. climate policy, transport policy)?

The chapter is structured as follows. Section two presents different concepts of sustainability indicators and discusses how sustainability criteria can be derived from them. Section three and four discuss the applicability of sustainability criteria to climate change and transport.

Sustainability Indicators

Sustainability: Definition, Rules, Indicator Concepts

Especially since the Rio United Nations Summit of 1992, sustainable development has become an essential agenda item for international environmental policy. According to the Brundtland Report of 1987, a development is sustainable when it 'meets the needs of the present without compromising the ability of future generations to meet their own needs' (WCED, 1987). This is an explicitly anthropocentric definition of sustainable development.

Operational definitions and indicators are a prerequisite for implementing sustainability in practical policy decisions. Although the general definition of sustainability touches upon nearly all areas of economic, ecological and social development, three main management rules of resource use have been derived from it (Daly, 1990):

- Harvest rates of renewable resources should not exceed regeneration rates.
- Waste emissions should not exceed the relevant assimilative capacities of ecosystems.
- Non-renewable resources should be exploited in a quasi-sustainable manner by limiting their rate of depletion to the rate of creation of renewable substitutes.

These three management rules characterise a sustainable use of natural resources. Hence, sustainability indicators should reflect how far the actual use of natural resources is away from this objective. Daly has used the metaphor of 'plimsoll lines' to describe this function of scale limits (Daly, 1992). According to Opschoor and Reijnders (1991), main steps in the process of deriving such indicators are:

- Identification of main elements of natural capital and their economic functions.
- Selection of the most important, endangered elements to be chosen for an indicator set. Although the degree of endangerment is not known exactly in all cases, such a scoping process seems to be reasonable for setting priorities.
- Setting of standards being oriented on the rules of sustainable resource use.
- Construction of indicators reflecting the actual condition of the environment in relation to the sustainability standards.

Essential criteria of the German Council of Environmental Quality (SRU, 1994) for the assessment of sustainability indicators are:

- They should refer the actual environmental quality to sustainability standards.
- Appropriate levels of space and time should be used from an ecological perspective.
- All ecological functions should be considered; and
- Impacts of structural changes should be integrated (e.g. impacts of land use changes).

Up to now, several concepts and sets of indicators have been presented. Broadly, the concepts can be divided into two underlying strategies:

The Economic Strategy Neo-classical economists identify the inefficient use of natural resources as the main reason for environmental problems. This inefficiency is caused by market failure, due to external effects. Thus, the economic strategy focuses on getting prices right. External costs are estimated by different methods and suggestions are made to internalise these costs. An example is the study on 'Externalities of Fuel Cycles' of the European Commission (CEPN *et al.*, 1994). Indicators derived from the economic strategy are measured in monetary units, like the 'sustainable income' or 'green GDP'. Monetary indicators can be characterised as indicators of weak sustainability because they assume that manufactured and natural capital are close substitutes.[1] This means that costs of environmental deterioration (e.g. forest damage) can be compensated by benefits from manufactured capital (e.g. income).

The Ecological Strategy The ecological strategy analyses the impacts of economic activities on ecological systems. This strategy tries to keep ecosystems intact by protecting natural abilities like ecological stability or ecological resilience. Indicators derived 'from the ecological strategy are measured in physical units. Examples of such indicators are critical loads quantifying depositions which may have significant negative impacts on ecosystems in the long run. Physical indicators quantifying thresholds of critical ecological functions can be characterised as indicators of strong sustainability because they deny the degree of substitution that weak sustainability assumes.

The following sections will describe some economic and ecological indicator concepts and identify sustainability criteria which can be derived from different indicator concepts.

Weak Sustainability Indicators

Examples of weak sustainability indicators are damage cost calculations (to be found in several studies of social costs), concepts of integrated environmental and economic accounting ('green GDP') or multidimensional socio-economic indices like the 'Index of Sustainable Economic Welfare' (ISEW). Damage cost calculations try to quantify the external effects of environmental pollution. The methodology is based on welfare theory and cost-benefit analysis. Two examples will be presented here, namely, the damage cost indicator of Pearce and Atkinson and the Huetings concept of sustainable income. The latter represents a stream of economists that are sceptical towards the possibility of calculating damage costs.

Both concepts, damage costs and sustainable income, are part of the System of integrated Economic and Environmental Accounts of the United Nations (SEEA).

Sustainability Indicator of Pearce and Atkinson Pearce and Atkinson have developed a sustainability indicator on the basis of damage cost calculations (Pearce and Atkinson, 1993). The sustainability criterion of their approach is that an economy should save more than the combined depreciation on natural and human-made capital. Estimates for savings and depreciation on both forms of capital are made for 22 countries (see Table 1). According to these estimates, developed countries like Germany or the United States are sustainable because of high savings ratios. Even East European economies like Poland and Hungary pass the sustainability test. Only developing countries like Ethiopia, Burkina Faso or Papua New Guinea fail the test. As Pearce and Atkinson remark, their weak sustainability indicator should be supplemented by strong sustainability approaches identifying critical elements of natural capita

Table 8.1   Sustainability of Selected National Economies

Sustainable Economies	S / Y -	δ_M / Y -	δ_N =	Z
Brazil	20	7	10	+3
Costa Rica	26	3	8	+15
Czechoslovakia	30	10	7	+13
Finland	28	15	2	+11
Germany (pre-unification)	26	12	4	+10
Hungary	26	10	5	+11
Japan	33	14	2	+17
Netherlands	25	10	1	+14
Poland	30	11	3	+16
USA	18	12	4	+2
Zimbabwe	24	10	5	+9
Marginally Sustainable				
Mexico	24	12	12	0
Philippines	15	11	4	0
Unsustainable				
Burkina Faso	2	1	10	-9
Ethiopia	3	1	9	-7
Indonesia	20	5	17	-2
Madagascar	8	1	16	-9
Malawi	8	7	4	-3
Mali	-4	4	6	-14
Nigeria	15	3	17	-5
Papua New Guinea	15	9	7	-1

Note: S = National Savings, Y = National Income, δ_M = depreciation on man-made capital, δ_N = depreciation and damage to natural resources and the environment (data for δ_N is taken from studies estimating monetary values of environmental damages for these countries).

Source: Pearce, D. W. and Atkinson, G. D. (1993), p.106

Hueting's Concept of Sustainable Income   Damage costs can often not be calculated because individual preferences are unknown and information about benefit losses connected with environmental degradation do not exist. The underlying assumption of Hueting's concept of sustainable income is that a societal consensus exists or should exist and that resources should be used in a sustainable manner, meaning that the functions of media like water, soil, forest and air should remain intact. On the basis of this consensus, sustainability standards are derived for the calculation of a sustainable income. Avoidance costs are estimated which have to be spent for achieving certain standards (see Figure 1). The ecological functions themselves are not valued in this approach (Hueting and Bosch, 1991).

The Hueting concept is justified under the implicit assumption that the (unknown) benefits attaining critical threshold levels of environmental functions are in excess of the costs of attaining those levels. However, hypothetical avoidance costs are not simple to estimate because long-term technical change must be forecasted for these calculations. Without a correct anticipation of technological progress, the real burden of achieving sustainability is overestimated. Estimations of avoidance costs differ significantly with regard to their aggregation levels. For example, many bottom-up studies identify substantial 'no regrets' potentials of climate change caused by market failure, while most top-down models do not include market failure and, consequently, calculate higher costs (IPCC, 1995).

The Hueting concept leads directly towards a paradigm of strong sustainability because it *assumes* that threshold values of ecological functions exist. The important question is, of course, *how* such threshold values can be derived. This question has to be answered with the help of physical indicators.

Figure 8.1 The Concept of Sustainable Income

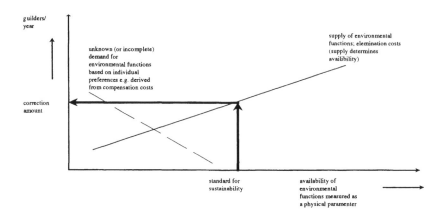

Source: Hueting/Bosch (1991), p.31

Strong Sustainability Indicators

In economic theory, strong sustainability concepts are especially discussed among ecological economists. While neo-classical economic theory is mainly interested in questions of allocation, e.g. correct prices, external effects and cost-benefit analysis, ecological economics claims to be transdisciplinary and open for new methodological approaches. An important target of ecological economics is the analysis of environmental problems beyond mere allocation questions. Following Daly, a clear separation should be made between the policy goals of allocation, distribution and scale (Daly, 1992). Scale refers to the ecological carrying capacity requiring that economic activities should not jeopardise the functions of ecosystems. The absolute carrying capacity of ecological systems should be protected as a prerequisite for dealing with distribution and allocation of natural resources. Processes which are relevant to resource use or to flows of matter-energy should be restricted according to a sustainable scale. The scale has to be measured in absolute physical limits.

Unfortunately, the concept of Daly in this context is based on the law of entropy which is vague and still disputed, having failed to provide evidence to explain interactions in open economic and ecological systems.

More evidence can be given to ecological approaches which are based on criteria like the resilience and functions of ecosystems, environmental viability or safe minimum standards. Although the process of specifying and quantifying these criteria evokes several problems, some progress has been made in developing physical sustainability indicators during the last years (Rennings, 1994 and 1999).

Ecocapacity   The Dutch Advisory Council for Research on Nature and Environment (RMNO) has developed ecocapacity indicators reflecting sustainability restrictions to resource use (Weterings and Opschoor, 1992). Expected environmental impacts of global demographic and economic developments until the year 2040 are related to acceptable sustainable impacts. Sustainability criteria have been derived for resource depletion, pollution and encroachment to quantify environmental targets. Ten macro-indicators have been roughly estimated. The main result is that necessary reductions due to the difference between expected environmental impacts and acceptable sustainable levels range between 20 per cent (for the depletion of coal) and 99 per cent (for the extinction of species) (see Table 2).

Material Intensity per unit Service   The German Wuppertal Institut has developed an indicator concept for measuring the Material Intensity of Products and Services (MIPS) reflecting the eco-efficiency of products (Schmidt-Bleek, 1994). The numerator informs about the materials use of products from 'cradle to grave' and is calculated in kilograms or tons. The denominator measures the services or benefits resulting from the material use, e.g. kilometres of travel. For example, the indicator can compare the material intensity of transport services for different traffic modes.

Materials use is an input measure of economic activities and does not provide information on the different environmental impacts resulting from this activity. Like energy efficiency, the indicator may be useful as a rough, macro-indicator of the technological progress towards sustainability. For the management of ecosystems at the micro-level, however, more specific information on emissions, structural changes and impacts, among others, is necessary. Schmidt-Bleek, the developer of MIPS, asserts that the material use in the industrialised countries has to be reduced by a factor of ten in order to achieve sustainability.

Critical Loads and Level   The main problem of environmental policy strategies based on best available technologies (BAT) is that they do not allow specific environmental impacts to be taken into consideration. The same problem is encountered by strategies of reducing emissions to a certain level. For example, all states signing the first SO_2-protocol were obliged to reduce their SO_2-emissions at least by 30 per cent. However, even such reductions could not stop the increasing acidification of forest soils. The recognition of the deficits of emission-oriented strategies has contributed to the development of the critical load and level concept of the Economic Commission of the United Nations for Europe (UN-ECE).

The critical load/level concept is the first approach using criteria for the environmental quality of complex ecosystems. It tries to derive critical depositions, especially for air pollutants, which may have significant negative impacts on ecosystems in the long run. In the definition of the UN-ECE, critical loads and levels are estimates of an exposure below which significant harmful effects on elements of the environment do not occur. The term critical loads is used for depositions, the term critical levels for the critical concentrations of these depositions (Nagel, Smiatek and Werner, 1994).

Table 8.2 Sustainable Versus Expected Level of Environmental Impact

Dimension/Indicator of Environmental Impact	Sustainable Level	Expected Level 2040	Desired Reduction	Scale
Depletion of Fossil Fuels				
Oil	Stock	stock	85 per cent	Global
Natural gas	For	Exhausted	70 per cent	Global
Coal	50 Years		20 per cent	Global
Depletion of Metals				
Aluminium	Stock	stock for > 50 years	None	Global
Copper	For	stock exhausted	80 per cent	Global
Uranium	50 Years	Depends on use nuclear energy	Not quantifiable	Global
Depletion of Renewables Resources:				
Biomass	20 per cent terr. animal biomass	50 per cent terr. animal biomass	60 per cent	Global
	20 per cent terr. Primary production	50 per cent terr. Primary production	60 per cent	Global
Diversity of species	Extinction 5 species per annum	365 – 65.000 species per annum	99 per cent	Global
Pollution:				
Emission of CO_2	2,6 Gigatonnes carbon per annum	13,0 Gigatonnes carbon per annum	80 per cent	Global
Acid deposition	400 acid eq. per hectare per annum	2400-3600 acid eq.	85 per cent	Continental
Deposition nutrients	P: 30 kg. per ha. per annum	no quantitative data	Not quanti-fiable	National
	N: 267 kg. per ha. per annum	no quantitative data	Not quanti-fiable	National
Deposition of Metals:				
Deposition of cadmium	2 tonnes per annum	50 tonnes per annum	95 per cent	National
Deposition of copper	70 tonnes per annum	830 tonnes per annum	90 per cent	National
Deposition of lead	58 tonnes per annum	700 tonnes per annum	90 per cent	National
Deposition of zinc	215 tonnes per annum	5190 tonnes per annum	95 per cent	National
Encroachment:				
Impairment through dehydration	Reference year 1950	no quantitative data	Not quanti-fiable	National
Soil loss through erosion	9,3 billion tonnes per annum	45 to 60 billion tonnes per annum	85 per cent	Global

Source: Weterings and Opschoor, 1992

Sustainability rules require that the assimilative capacity of ecosystems is not jeopardised. This demand can be quantified in an operational way by using critical loads and levels (Gregor, 1995). The latter indicate the degree to which emissions have to be reduced to keep environmental impacts within an acceptable limit. However, environmental standards are often political targets which are not oriented toward critical levels and loads, especially not in the short run.

The process of deriving critical loads/levels can be divided into four phases (Gregor, 1995):

• Deriving benchmarks for emission impacts from analyses of ecosystems and experiments;
• Discussion and finding consensus in the scientific community;
• Characterising the impacts by mapping critical loads (mapping-programme);
• Discussion of reduction targets to keep depositions under a critical level.

Critical levels are derived from experiments in laboratories and in the field. Criteria for deriving critical loads and levels include (Nagel, Smiatek, and Werner, 1994): the time lag until certain concentrations lead to damages; synergistic and additive effects of substances; external factors like temperature having influence on the resistance of receptors; and internal factors, like the state of the development of organisms.

In Germany, SO_2-benchmarks could be differentiated for certain receptors as forests or different plants. For NO_2 and O_3 only general values could be derived. The maps of critical levels for Germany show that: for SO_2 the critical levels were exceeded on about half of the area of forests and natural vegetation in 1988 and on about one third in 1990 (see Figure 2); for NO_2 the critical levels were exceeded on about one third of the whole area in Germany in 1988 and 1990; for O_3 the critical levels were exceeded on about 90 per cent of the whole area in Germany in 1988 and 1990 (average values per month) (Nagel, Smiatek and Werner, 1994).

Critical loads in Germany are derived for the acidification of soils as an aggregation of the effects of different acid depositions (SO_2, NO_2). Main receptors being analysed are woods and, to a certain degree, aquatic ecosystems. Maps of German forest areas show that in more than 85 per cent of the area, critical loads – quantified on the basis of a mass balance[2] – are exceeded.

Compared with the lack of refinement of other approaches, the critical load concept is methodologically well-founded. Thus, the German Council of Environmental Advisors (SRU) has requested further research in this field. Beyond that, the SRU suggested developing indicators for critical structural changes not related to certain material uses, for instance impacts on certain habitats caused by land use changes (SRU, 1994).

Pressure-State-Response Indicators The Pressure-State-Response (PSR) approach of the OECD has been derived from the Canadian Stress approach. Indicators are classified in three categories (Scherp, 1994):

- Pressure indicators try to answer questions about the causes. Indicators in this category include emissions and waste amounts.
- State indicators answer questions about the state of the environment. Indicators in this category include urban air quality, ground water quality, temperature changes, concentrations of toxic substances or the number of endangered species.
- Response indicators try to answer questions about what is being done to solve the problem. Indicators placed in this category include international commitments or recycling rates.

The aim of the OECD indicators is to allow international comparisons of environmental indicators. Thus, the process of indicator selection has to be oriented towards the availability of data in all member countries. Hence, the indicator set is not ecologically well-founded. Indicators are not related to sustainability goals and give little information about essential functions and structures of ecosystems. Nevertheless, the OECD system can be used as a first step to implement more advanced indicator sets in the future.

Driving Force-State-Response Indicators More recent concepts have tried to integrate social and economic components into indicator systems. Based on the PSR concept, the UN Commission on Sustainable Development (CSD) has developed the 'Driving Force-State-Response' approach (DSR) (United Nations, 1995). Additionally, the DSR introduces the category of institutional indicators since institutional change has been recognised as an important area of sustainable development in Agenda 21. The DSR includes the following indicator categories:

- Driving force (social, economic, ecological development).
- State (state concerning sustainability).
- Response (policy reactions and options for sustainable development).

The working programme of the CSD for developing sustainability indicator systems emphasises the importance of integrated indicators reflecting 'inter-linkages' between social, economic and ecological development. Priority is also given to highly aggregated indicators. The working programme is based on a preliminary menu of indicators (see Appendix). It is planned to use the menu in a flexible manner considering national peculiarities and new scientific insights or experiences.

Eurostat (1997) has published DSR indicators as far as they are available for the European Union. The economic DSR indicators are shown in Table 3; those available at Eurostat for the European Union are indicated in bold.

The suggested menu is oriented on the chapters of the Agenda 21, thus DSR-indicators can be characterised as pragmatic indicators for monitoring progress on issues raised by the Agenda. Problems remain concerning the relevance and meaning of some indicators for different countries (e.g. relevance of indicator 'mega-cities with 10 million or more inhabitants' for most European countries) and the consistency of the whole system (e.g. the highly aggregated indicators 'environmentally adjusted GDP' is listed as a state-indicator in the category 'international co-operation', together with indicators like 'export concentration ratio'). Indicators are furthermore missing for some chapters of the Agenda 21 (e.g. consumption patterns or decision-making structures), whereas the usefulness of institutional indicators is unclear.

All in all the menu can be described as 'quick and dirty'. It is nevertheless useful as a first pragmatic step towards sustainability indicators. It is also politically highly relevant since no single approach can consider all relevant economic, social and ecological aspects for sustainability indicator systems in a consistent way.

From Sustainability Indicators to Sustainability Criteria

The use of a certain indicator concept determines both the corresponding sustainability criteria and the evaluation method.

Example 1: Ecocapacity Indicators The underlying concept of the eco-capacity indicators is the idea of environmental space. It follows the normative assumption that limited environmental resources and services should be distributed equally. Thus the calculated reduction targets are based on the central sustainability criterion 'equal intra- and intergenerational share of environmental services'.

An appropriate evaluation method would be to use critical thresholds as far as possible for environmental resources and services, and subsequently cost-effectiveness analysis to identify least-cost-options for achieving sustainability targets.

Example 2: Material Intensity The underlying normative assumption is that material use in industrialised countries must be reduced by a factor of ten for achieving sustainability. In this context material intensity is a measure for progress towards this long-term sustainability target. Improving eco-efficiency is the basic sustainability criterion under this approach. Economic aspects are addressed by expressing the belief that synergies between economic and ecological objectives exist. Thus economic studies of proponents of this approach normally try to identify win-win-options and no-regrets-potentials.

Example 3: Critical Loads and Levels The basic sustainability criterion is that critical levels and loads should not be exceeded. Existing models of integrated economic and ecological assessment calculate least cost options for reaching these goals. Thus cost effectiveness analysis is used for addressing the economic dimension of sustainability.

Example 4: PSR and DSR Approach The PSR and DSR indicators avoid to use normative assumptions. They are open concepts for the classification and description of environmental, social and economic trends, and can be characterised as performance indicators. Every single indicator can be used as a criterion for sustainability, but no target values are given and the indicators are not weighted. Thus evaluation methods are necessary to derive sustainability criteria. For instance, multi-criteria analysis can be used for the weighting of indicators, cost-benefit analysis for the quantification of peoples' willingness to pay, and Delphi methods for expert judgements.

Table 8.3   Economic sustainability indicators of the Commission on Sustainable Development

Chapters of Agenda 21	Driving Force Indicators	State Indicators	Response Indicators
Chapter 2: International cooperation to accelerate sustainable development in countries and related domestic policies	GDP per capita Net investment share in GDP Sum of exports and imports as a per cent of GDP	Environmentally adjusted net domestic product Share of manufacturing value-added in GDP (exports)	
Chapter 4: Changing consumption patterns	Annual energy consumption per capita Share of natural resource intensive industries in manufacturing value-added	Proven mineral reserves Proven fossil fuel energy reserves Lifetime of proven energy reserves Intensity of material use Share of manufacturing value-added in GDP Share of consumption of renewable resources in GDP Consumption of renewable energy	
Chapter 33: Financial resources and mechanisms	net resources transfer/GND Total Official Development Assistance (ODA) given or received as a in per cent of GNP	Depth/GNP Dept service/export	Environmental protection expenditures as a percent of GDP Amount of new or additional funding for sustainable development
Chapter 34: Transfer of environmentally sound technology, co-operation and capacity-building	Capital goods imports Foreign direct investments	Share of environmentally sound capital goods import	Technical co-operation grants

Source: United Nations (1995), Eurostat (1997)

The Case of Global Warming

In the Kyoto Protocol some of the Parties to the Framework Convention on Climate Change commit themselves to reduction targets for the different greenhouse gases. In accordance with Article 3.1 of the Convention, targets for individual states should be defined 'in accordance with their common but differentiated responsibilities and respective capabilities'.

The role of scientific research in this process is to help decision makers to derive reasonable reduction targets. However, the appropriate decision rule depends on the underlying interpretation of the sustainability paradigm which can follow an ecological or an economic approach.

- From a welfare theoretical perspective, a (weak) sustainability approach is based on the principle that social welfare should be maximised and the total costs of climate change (abatement, adaptation and damage costs) should be minimised. The idea of the weak sustainability criterion is that any environmental damage should 'be compensated by projects specifically designed to improve the environment' (Markandya and Pearce, 1991). This sustainability criterion is 'weak' because it allows for unconstrained elasticity of substitution between different types of natural capital.
- From an ecological perspective, the (strong) sustainability rule requires that the total sum of greenhouse gas emissions should not exceed the assimilative capacity of the atmosphere and that, at least, irreversible and catastrophic effects on the global ecosystem should be avoided. In this sense strong sustainability regards natural capital as providing some functions that are not substitutable by man-made capital. These functions, labelled 'critical natural capital', are stressed by defining sustainability as leaving the future generations a stock of natural capital not smaller than the one enjoyed by the present generation (Cabeza Gutés, 1996).[3]

In several contributions, damage cost calculations of climate change like that of Nordhaus (1991) and Fankhauser (1995) were criticised, especially from an ecological perspective. It has been argued that mere neo-classical optimisation concepts tend to ignore the ecological, ethical and social dimension of the greenhouse effect, especially issues of an equitable distribution and a sustainable use of non-substitutable, essential functions of ecosystems.[4]

The ecological argument criticises the use of damage cost values for computing optimal levels of emission abatement as neglecting the special function of the atmosphere as a sink for greenhouse gases. This function is absolutely scarce and essential for the global ecosystem. It is feared that by putting certain monetary values on this essential natural function politicians may be encouraged to 'sell' this in exchange for goods of 'higher' value in the short-time horizon (e.g. income).

The following sections will discuss the weak and strong sustainability approaches to climate protection and show their applications, weaknesses, possible improvements and inter-linkages. First, we discuss damage cost studies of global warming as representative of the weak sustainability approach. Since a detailed description of the contents and results of existing studies has already been done in the IPCC Second Assessment Report (Pearce *et al.*, 1996), we instead focus on the critical issues of the studies and the further development of their methodological framework. Subsequently the 'inverse scenario' approach of the German Advisory Council on Global Change (WBGU) and the 'environmental space concept' of the Dutch Advisory Council for Research on Nature and Environment (RMNO) are used to illustrate operational indicators of strong sustainability. Finally, the role of cost-benefit analysis for the economic assessment of climate change is discussed.

Weak Sustainability Indicators of Climate Change

Major valuation studies undertaking an estimation of external costs in the energy sector renounce the use of damage cost values due to empirical and methodological problems. This is the case of the valuation studies of the European Commission (ExternE) and of the U.S. Department of Energy (DOE Study). One of the authors of the DOE Study states: 'The earlier studies include estimates of damages from climate change; the more recent studies do not include them in their summary tabulations' (Lee, 1996). In a footnote the same author remarks that 'this conclusion does not say that damages from climate change are zero, but that precise estimates of these damages do not have a sound scientific basis because of great uncertainty'.

An alternative strategy has been the calculation of abatement costs (for specific CO_2-reduction targets) instead of damage costs. Most advocates of an ecological paradigm of sustainable development prefer the use of abatement costs because they are normally related to CO_2-reduction targets leading to sustainable future emission paths. The abatement cost option has been chosen by studies from de Boer and Bosch (1995), Bernow *et al.* (1996), and Ott (1996).

A review of the state-of-the-art concerning the economic valuation of damages, including the major studies mentioned above, has been undertaken by the Intergovernmental Panel of Climate Change (IPCC) in it's Second Assessment Report.[5] The report cites the range of estimates of marginal damage at 5-125 $ per ton of emitted carbon (Pearce *et al.*, 1996). Working Group III of the IPCC has given special attention to the role of cost-benefit analysis and the incorporation of intra- and intergenerational equity aspects. It identified key problems that are not adequately addressed by the use of traditional cost-benefit analysis in the assessment of climate change (IPCC 1995, WG III; Arrow, Parikh, Pillet *et al.*, 1996). These include large uncertainties, long time horizons, the global, regional and intergenerational nature of the problem, wide variations of the cost estimates of potential physical damages due to climate change, wide variations of the cost estimates of mitigation options, low confidence in monetary estimates for non-market impacts, possible catastrophes with very small probabilities and issues of intragenerational equity (especially lower values for the statistical lives of people in developing countries as compared to those in developed countries).

Additionally, there are serious ethical concerns about the normative assumptions of external cost studies. These are often not explicit. One assumption is that economic welfare is measured by people's willingness to pay or their willingness to accept compensation. Consequently, the welfare of rich people or of rich nations carries more weight in the results than the welfare of poor people or nations. Especially disputed is the widespread method of valuing human lives in developing countries as lower than those in developed countries. Another implicit, albeit central, judgement concerns the possibility of compensating future individuals for climate damages. Such assumptions have to be made transparent.

Responding to the IPCC criticisms, Fankhauser and Tol (1995) and Tol (1996b) derived a research agenda for the economic assessment of climate change impacts. This includes improved damage estimates for less developed countries; improved estimates for non-market losses, especially morbidity and ecosystem effects; an assessment of the importance of variability and extreme events; models of the process of adaptation and the dynamics of vulnerability; formal uncertainty assessments and analyses of the outcomes; an improved comparison and aggregation of estimates between countries and between generations; and guidelines ensuring consistency between economic and non-economic impact assessment.

Following this agenda, first progress can be observed, especially with regard to the handling of intra- and intertemporal equity.[6] Intragenerational equity has been addressed by Fankhauser, Tol and Pearce (1996) as well as by Azar and Sterner (1996). Both use an approach of equity weighting: based on existing estimates of global warming damages, willingness to pay values are adjusted in the aggregation process. While aggregating estimates for single countries or world regions to a global value, the damages are weighted by the inverse of income. Damages of rich countries are weighted down and damages of poor countries are weighted up by adjusting these damages to the average annual per capita world income. The reason for the adjustment is 'decreasing marginal utility of money and for the same reason we can argue that a given (say one dollar) cost which affects a poor person (in a poor country) should be valued as a higher welfare cost than an equivalent cost affecting an average OECD citizen' (Azar and Sterner, 1996). Using equity weighting, damages and deaths in developed countries do not count more than in developing countries. The annual world income is used as a budget restriction.

Advances have also been made in the field of measurement of intergenerational equity. The results of monetary values of climate change damages depend substantially on the choice of the discount rate. The higher the discount rate, the lower the present value of future damages. Thus, discounting is often criticised because it produces incentives to shift environmental risks from the present to the future. However, the relationship between discount rate and climate change is very ambiguous. Lowering the discount rate induces an increasing level of economic activity and investment. This would probably lead to further emissions of greenhouse gases (CEC/US Joint Study, 1993). The relationship between the discount rate and environmental deterioration is known as the 'conservationist's dilemma', since both high and low discount rates can favour environmental conservation (Norgaard and Howarth, 1991).

Cost-benefit analyses tend to use a range of discount rates. Following Markandya, discount rates of 0.3 to ten per cent represent an adequate range of parameters for the European Union (CEC/US Joint study, 1993). Three per cent are taken as a rate for social time preference, zero per cent and ten per cent as extreme parameters for sensitivity analysis.[7] With regard to climate change, none of the three rates is satisfying: while rates of three to ten per cent lead to nearly zero costs for long term damages, a rate of zero per cent may evoke infinite costs.

The rate of three percent is derived from the concept of social time preference (STP), a measure of the decline of social welfare or utility of consumption over time (Markandya and Pearce, 1991). The social time preference depends on the rate of pure individual time preference (ITP) or impatience, on the growth rate of real consumption per capita (W), and on the elasticity of the marginal utility of consumption (U). The equation is:

$$STP = ITP + W \times U \tag{1}$$

An argument against the STP concept is that ecological limits to (economic) growth will set biophysical constraints on real consumption per capita (W) in the long run. Such constraints should be taken into consideration. For the EU, Markandya recommends a rate for W of around one or two per cent (CEC/US Joint Study, 1993).

From an environmental perspective it is argued that the individual time preference (ITP) should be refused in social investment decisions. This position takes the perspective of society as a whole and criticises impatience as being irrational. Contrary to the individual view, for a society it would seem unreasonable to privilege present preferences above future preferences. However, a collective view conflicts with methodological individualism being a fundamental element of welfare economics.

Rabl (1993) argues that a discount rate for intergenerational effects should be defined by taking the perspective of future generations. From Rabl's point of view, market interest rates can only be considered to the extent that a market exists. Following Rabl, the longest time horizon of market transactions is 30 to 40 years. Thus, there is no inconsistency in lowering the interest rate for damages beyond that time horizon. In consequence, Rabl recommends to split the social time preference (STP) in an:

$$STP = ITP + W \times U \qquad \text{for short term effects (< 30 to 40 years)} \tag{2}$$

and an

$$STP = W \times U \qquad \text{for long term effects (> 30 to 40 years)} \tag{3}$$

At first glance, the splitting concept and the time horizon for market transactions chosen by Rabl seem to be very arbitrary. Upon closer inspection, a special treatment of long term effects seems to be reasonable, because otherwise damages occurring a hundred or more years in the future will be totally ignored in monetary valuation studies. Understood as a first rule of thumb, Rabl's concept of time-variant discount rates helps to improve the treatment of long term effects in external cost studies.

More accurate values can be estimated by using models of overlapping generations (OLG-models). For example, Bayer and Cansier (1996) have developed a simple OLG-model including four generations with a life expectancy of four periods for each generation. A more realistic, but complex model would include about 40 generations. The aim of OLG-models for calculating costs of climate damages is to estimate the discounted value of investments in climate protection when benefits go beyond the life expectancy of the current generation. The calculation is done year by year considering the demographic structure of the current generation. While all effects of climate protection on consumption within the life expectancy of the current generation are discounted by using social time preference including individual time preference, all effects beyond the current generation are discounted by using social time preference without individual time preference.

It can be summarised that the introduction of time-variant discount rates is reasonable with regard to long-term environmental damages. To express it in the words of Sterner and Azar (1996), 'a constant discount rate should only be seen as a special case of the more general case where the discount rate is allowed to vary'.

A more radical position is to set the discount rate equal to zero (Pearce, 1993). A zero discount rate follows the rule that consumption at one point of time does not count more than welfare at another point of time. However, it is feared that a zero discount rate would imply infinite social costs and total current sacrifice (Pearce, 1993).

Nonetheless, there are good reasons to use zero discount rates for certain natural resources (W_{NR}), the market values of which are expected to rise proportionally to the Gross Global Product (GGP). One important example is the demand for safety, expressed in terms of the statistical value of life (VSL). It can be assumed that the growth rate of peoples willingness to pay for reducing health risks will be at least as high as the growth rate of the Gross Global Product. All things considered, zero per cent seems to be an appropriate discount rate only if growth rates can be expected to be zero or if growth rates of people's willingness to pay for a reduction of health risks can be expected to be higher than the average growth rate of the economy.

One controversial question with regard to discounting concepts is that of compensation between generations. The use of market interest rates assumes compensation from one generation to another for losses of natural capital; the use of lower discount rates assumes that environmental protection is the only way to make these intragenerational transfers (Arrow, Cline, Mäler *et al.*, 1996).

Intergenerational fairness is concerned with distribution across generations. It would seem reasonable to separate this issue of distribution from issues of efficiency. According to Daly, 'the policy instrument for bringing about a more just distribution is transfers – taxes and welfare payments' (Daly, 1992). Norgaard and Howarth argue in the same direction by pointing out that

> if we are concerned about the distribution of welfare across generations, then we should transfer wealth, not engage in inefficient investments. Transfer mechanisms might include setting aside natural resources, and protecting environments, educating the young, and developing technologies for the sustainable management of renewable resources. Some of these might be viewed as worthwhile investments on the part of this generation, but if their intent is to function as transfers, then they should not be evaluated as investments. The benefits from transfers, in short, should not be discounted (Norgaard and Howarth, 1991).

From this point of view, discounting should only function as a mechanism for efficient allocation. Distributional aspects should be treated separately from allocation, although they are not independent. It is plausible that transfers to future generations change relative prices. As Norgaard and Howarth remark:

> With different distributions and efficient allocations, new prices arise. One can no more speak of 'the' rate of interest when societies are giving major consideration to the sustainability of development than one can speak of 'the' price of timber when deciding whether to conserve forests. Redistributions change equilibrium prices (Norgaard and Howarth, 1991).

It can be concluded that the approach of deriving weak sustainability indicators of global warming by estimating damage costs requires strong normative choices about intra- and intergenerational fairness and the handling of uncertainty. These normative choices are made in most cases implicitly, i.e. they are hidden under a veil of aggregation and discounting rules. These problems have been especially emphasised by the Second IPCC Assessment Report. However, important responses to the IPCC criticism have been made. While the long-term dynamic effects of global warming and the resulting social and economic impacts are still not well understood, some important contributions have been made with regard to an improved handling of intra- and intergenerational equity issues. To summarise, in so far as issues of allocation are concerned, an appropriate range of discount rates should integrate

* Zero per cent as a rate for long-term effects which are expected to rise with GDP;
* One per cent as rate for social time preference (STP) ignoring individual time preference (ITP);
* Three per cent as a rate for social time preference (STP) including individual time preference (ITP); and
* Higher discount rates representing market interest rates (concept of opportunity costs of capital).

The concept of time-variant discount rates seems to be consistent with the principles of welfare theory. While three per cent can be used as a standard discount rate, lower rates can be applied for the long-term global warming effects.

Equity weighting and time-variant discounting can determine which degree of investments for stabilising the global temperature can be justified economically. A first application of time-variant discount rates and equity weighting as described above suggests that the marginal damage costs for carbon range from 260 to 590 $ per ton, which is 50 to 100 times higher than the Nordhaus value of 5 to 125 $ cited in the IPCC report (Azar and Sterner, 1996).

Strong Sustainability Indicators of Climate Change

Environmental Space Important early contributions discussing acceptable levels of greenhouse gas emissions were made by the Dutch Advisory Council for Research on Nature and Environment (RMNO) (Weterings and Opschoor, 1994) and by the German Enquete Commission 'Preventive Measures to Protect the Earth's Atmosphere'. The Enquete Commission derived specific national reduction targets and general targets for developed and developing countries following the recommendations of the 1988 World Conference on Atmospheric Change in Toronto (German Bundestag, 1991). The Toronto conference had recommended the reduction of global emissions of CO_2 by about 20 per cent relative to the 1988 emission levels by the year 2005, and their subsequent reduction by over 50 per cent by the year 2050 (German Bundestag, 1991).

The recommendations of the RMNO and the Enquete Commission are based on a study of Krause, Bach and Koomey (1990) that estimated tolerable CO_2-emissions. They calculated a tolerable relative deviation of 0.1°C per decade (data on the ability of trees to migrate suggest this as a maximum rate of temperature-rise) and an absolute warming limit of 2.0-2.5°C above pre-industrial level for the next 100 years (which would lead to a maximum acceptable sea level rise of one meter in the forthcoming centuries). The authors assumed that the most important ecological functions could sustain such a temperature change. These thresholds were subsequently translated into critical concentrations and critical emission paths.

In a report to the RMNO, Weterings and Opschoor (1994) used the concept of environmental space to share the global budget for CO_2-emissions among nations. They describe the concept of environmental space as follows:

> Environmental utilisation space [or environmental space] is a concept which reflects that at any given point in time, there are limits to the amount of environmental pressures that the earth's ecosystem can handle without irreversible damage to these systems or to the life support processes that they enable. This suggests to search for the threshold levels beyond which actual environmental systems might become damaged in the sense indicated above, and to regard this set of deductively determined critical values as the operational boundaries of the environmental space (Weterings and Opschoor, 1994).

Five criteria were used for distributing the carbon-budget over regions and nations, namely, GNP, land area, current energy consumption (status quo criterion), current population (equity current criterion; equal emission per capita), as well as current and future population (equity cumulative criterion; equal emission per capita).

Table 8.4   Sustainability Criteria for OECD-Carbon Release

Criterion	OECD Budget 1985-2100 (% global budget)	(GtC)	OECD Annual average (GtC)	% current emission
GNP	63 %	189	1.64	57 %
Land area	24 %	72	0.63	22 %
Status Quo	47 %	140	1.17	42 %
Equity current	16 %	48	0.42	15 %
Equity cumulative	11 %	33	0.29	10 %

(Assumption: Sustainable world carbon budget 300 GtC as estimated by Krause *et al.* 1990; OECD current annual release 2.8 GtC)

Source: Van der Loo (1993)

Following these five criteria, the global carbon-budget was distributed among nations and regions. The different sustainability indicators (according to different distribution criteria) were compared to the actual and forecasted performance of the OECD countries (see Table 8.4). The report concluded that

> the OECD does not meet the various sustainability criteria currently and is not forecasted to do so in the forthcoming decades. Nor does any of the individual Member States. Even if we forget about a more equal distribution in respect of the developing countries, the OECD emission exceeds sustainability (status quo) by more than a factor 2. From the equity perspective the OECD performance is unsustainable by a factor of 7 to 10 (van der Loo, 1993).

The definition of strong sustainability standards requires some normative choices. Compared with 'weak' optimisation concepts, the advantage is that these choices are made explicit, i.e. that they are not hidden under a veil of aggregation and discounting rules.

The 'Backcasting Scenario' or 'Tolerable Window Approach' Another example of strong sustainability indicators of global warming is the 'backcasting scenario' of the German Advisory Council on Global Change (WBGU). Based on this, the WBGU draws the conclusion that the acceptable absolute positive deviation from the present mean temperature on earth is 1.3 °C and a temperature change of 0.2 °C per decade is the upper tolerable limit.

The new scenario for the derivation of global CO_2-reduction targets and implementation strategies was published upon the occasion of the Climate Conference in Berlin (WBGU, 1995a). The scenario specifies tolerable impacts for humans and nature and subsequently, using the backcasting method, derives long-term global reduction targets (WBGU, 1995b). The backcasting scenario approach allows the adoption of strong sustainability criteria based on acceptable impacts or minimum standards of climate stability. It contains six steps (see Figure 8.2):

- In step one, a range of tolerable impacts caused by climate change is defined. Identifying tolerable impacts and damages, the 'backcasting scenario' starts explicitly with a *normative judgement.*
- Temperature changes are derived assuring that tolerable stresses are not exceeded.
- In step three and four, admissible concentrations and
- emissions of greenhouse gases (here only CO_2) are quantified by using models of climate dynamics and the carbon cycle.
- Next, the total emission reduction is broken down to states or regions.
- Finally, an efficient policy-mix for mitigating climate change is derived.

The basic normative principles of this approach are the preservation of Creation and the prevention of excessive costs. The principle of the preservation of Creation is formalised in the form of a tolerable 'temperature window' (WBGU, 1995b) derived from the natural temperature fluctuation during the geological period having shaped our present environment (late Quarternary period). The minimum and maximum values of this temperature window are the last ice age (10.4 °C) and the last interglacial period (16.01 °C). By extending this temperature range by 0.5 °C at either end, the window extends from 9.9 °C to 16.6 °C. Using these thresholds, the acceptable absolute positive deviation from the present mean temperature on earth (15.3 °C) is only 1.3 °C.

The principle of the prevention of excessive costs is defined very crudely in terms of losses of Gross Global Product (GGP). On the assumption that a disruption of economic systems will take place if losses of GGP exceed five per cent, this value is taken as a threshold for economic impacts. The possible unequal spatial distribution of damages across nations (e.g. for island states) and non-monetary burdens are not yet considered in this minimum standard. Most monetary estimates of doubling CO_2 concentrations until 2100 (mean temperature increase of roughly 0.2°C per decade) have calculated GGP losses of around one to two per cent. Considering that these calculations did not include several damage categories (e.g. extreme events) and may have underestimated the total costs, the WBGU sees 'good reason to assume that with a temperature change of 0.2 °C per decade the upper limit for adaptation costs of 5 per cent of GGP would be reached' (WBGU, 1995b).

The thresholds of the temperature window have been formulated as minimum standards for political reasons because the results should not be assessed too pessimistically. With the help of these operational criteria, a two-dimensional climate window is defined that should not be exceeded.

It is important to mention that the missing link of both backcasting and forecasting scenarios is the assessment of social and economic impacts. Pursuing a closed circle of integrated assessment, an economic analysis of different abatement and adaptation strategies would be desirable, including a valuation of remaining damages with monetary or non-monetary values.

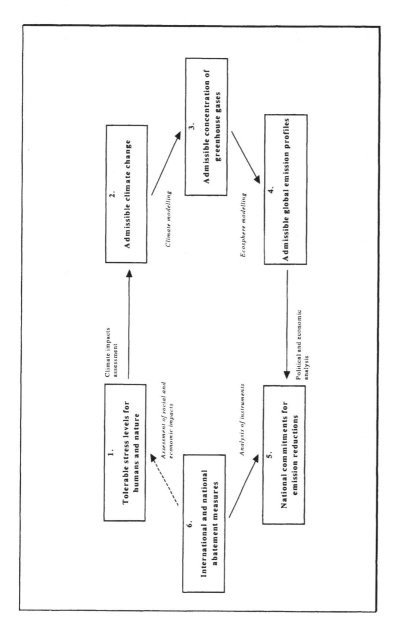

Figure 8.2   The Council's 'Inverse Scenario'

Sustainability Criteria and Cost-Benefit Analysis

It has been shown that the outcomes of studies estimating economic and social impacts in monetary units are very uncertain. However, as the alternative approaches of strong sustainability have illustrated, uncertainties do not disappear when norms are used instead. Schellnhuber, one of the developers of the 'inverse scenario', states that norms can only induce maximum or minimum values (for instance a safe minimum standard) but not optimums (Schellnhuber, 1995). If the identification of optimal emission paths between minimum and maximum standards were to be pursued, the strong sustainability approach would have to be supplemented by an economic impact assessment of damages, adaptation and abatement strategies.

Given perfect information about damage paths as well as present and future preferences, impact assessment would be able to replace the normative judgements of the 'inverse scenario'. However, in the light of decision-making under uncertainty, it becomes clear that a complete substitution of normative judgement by cost-benefit analysis or integrated assessment models will hardly be possible. Damage cost valuation techniques themselves contain central normative assumptions.

On the other hand, even normative target approaches often depend on monetary values for defining tolerable stress levels. The 'inverse scenario' documents this close link between acceptable emission paths and economic damages. Thus it becomes evident that further information about the global distribution of costs and benefits of climate change is desirable for the political negotiation process. For example, what is to be done when global average damages do not exceed five per cent of the gross global product, yet reach 100 per cent of the national income of certain island states and coastal zones? And how are high disparities of damages between economic sectors, social groups or species to be handled? Which damages can be compensated, which can not?

Many of these questions can only be answered by following a broader approach of strong sustainability including damage figures in so far as valid estimates are available. Tolerable stress levels can only be determined through negotiation as they vary across regions, economic sectors or societal groups. Thus, the relevance of imperatives like the preservation of creation and the prevention of excessive costs can be improved by more disaggregated sectoral and regional information about climate change impacts.

Therefore it seems reasonable to improve economic impact assessment and include it in integrated models for assessing climate change policy (WBGU, 1995b). Weak and strong sustainability indicators can be used in a complementary fashion in the assessment of climate change. Both can be understood as parts of a broader approach to integrated assessment modelling.

The critical IPCC review of social cost studies has influenced research and enhanced methodological progress, especially the handling of intra- and intergenerational equity issues. Further progress may lead towards more dynamic models and a multidimensional valuation of impacts. It should be noted that damage cost valuation is only one of several problems of cost-benefit analysis (Munasinghe *et al.*, 1996). According to the IPCC report, cost-benefit analysis should encompass a family of decision-analysis techniques like multi-criteria analysis or decision analysis. Using such a broad interpretation, social and economic impacts do not necessarily have to be described in monetary units.

A main disadvantage of cost-benefit analysis is that in complex decision situations relevant multiple criteria (e.g. efficiency, equity, uniqueness of resources or health and safety) are mixed and reduced to one single criterion. multi-criteria analysis may be a better way to show trade-offs between these different policy goals (Munasinghe *et al.*, 1996). An appropriate use of cost benefit techniques should thus be combined with a move towards a more frequent use of multiple, monetary and non-monetary valuation schemes.

The Case of Environmentally Sustainable Transport

Sustainability and Transport

Most studies on sustainable mobility deal primarily or mainly with passenger transport. The sustainable transportation principles elaborated by the 1996 Vancouver OECD Conference (see also next section) likewise focus mainly on passenger transport. Yet the rapid growth of freight transport and especially of road freight transport gives additional cause for concern.

In order to specify sustainability criteria for transport, it is first necessary to specify in which ways the contemporary European transport system is unsustainable. The following three questions must be answered:

- How unsustainable is transport today?
- What are the contemporary transport trends?
- Which problems and bottlenecks are likely to be faced on the path towards a more environmentally sustainable transport system?

Transportation is presently unsustainable for at least three reasons: First, it constitutes the major use of a non-renewable resource, namely oil; renewable substitutes for oil are at present not being developed at a sufficient rate. Second, it is responsible for 25 per cent of global CO_2 emissions from fossil fuel use. The tremendous growth of these emissions has contributed to a significant rise in atmospheric CO_2 concentrations as compared to their pre-industrial levels. Both passenger and freight transport tend towards emissions that exceed the assimilative capacity of the environment. Third, many of the local and regional impacts of transport activity damage the health of humans and other organisms and the integrity of ecosystems. At the local level, transport activity might be the most important contributor to health risks associated with toxic air pollutants (e.g. carbon monoxide, benzene and other volatile organic compounds, fine particulate matter and lead). At the regional level, the combustion of fossil fuels produces pollutants which can travel over large distances and which damage human health, plants, animals and ecosystems. These pollutants and their derivatives, such as tropospheric ozone and acidifying compounds, lead to a rise in respiratory problems and diseases among humans and impair the growth of crops and forests (OECD Environment Directorate, 1999).

The German Federal Environmental Authority (1997) adds that today's noise pollution constitutes a threat to health – contributing to high blood pressure and cardiovascular diseases – and must be brought down to permissible levels. Reductions are necessary to prevent adverse effects, like interference with communication and sleep disturbance. Some 130 million people in the EU are affected by noise emissions from road, rail and air traffic which exceed permissible levels, with 90 per cent of this being attributable to road noise alone.

The stress to which nature and landscapes are exposed to today through the continuous expansion of transport infrastructure, as well as the traffic-related changes in urban living conditions, are incompatible with sustainable living in general and with sustainable urban living in particular.

Important urban spaces are occupied by cars (...) children can hardly play or otherwise spend their time on or near the streets, elderly people feel insecure and confined when walking on the streets, and inner-city streets have virtually lost their function as a place to spend time and communicate with others (Umweltbundesamt, 1997).

The transport sector is additionally a major user of non-renewable resources and is closely related to the problem of waste and waste disposal. Indicative are the dimensions of material flows associated with the production, maintenance and disposal of motor vehicles. It has been estimated that the production of new cars in Germany alone consumes yearly 2.4 million tonnes of iron/steel, 300 k-tonnes of other metals, 450 k-tonnes of synthetic materials, 75 k-tonnes of glass and 500 k-tonnes of other materials (Umweltbundesamt, 1997). These amounts of material flows are not considered sustainable.

According to the 'business-as-usual' scenario of the OECD Environmentally Sustainable Transport (EST) project (OECD, 1999), the trends till 2030 are equally unsustainable, even if all present and planned legislative, technological and societal changes and measures were to be fully implemented.

Car ownership and total distance travelled will be at substantially higher levels in 2030 despite the fact that vehicles are expected to be more fuel efficient and less polluting. Gasoline and diesel will continue to be the most widespread source of transport energy, with some increase in the use of gas, hybrid, fuel cells and electric vehicles. Both light and heavy-duty road freight transport are likely to continue to increase and increases in road freight activity will be generally higher than those for car use. Rail and waterborne freight will also grow, but at a much lower rate. The biggest growth rates are forecasted for air transport and for both passenger and freight.

Transport noise nuisance will overall decrease slightly, whereby the noise associated with aircraft is expected to increase. Emissions of nitrogen oxides, carbon monoxide, volatile organic compounds and fine particulates per vehicle-kilometre will decrease substantially from their 1990 levels, yet caution is required as the growth in volume of travel might inverse this positive trend. What however will not decrease, but rather increase, are carbon dioxide emissions from the transport sector. These are forecasted to double by 2030 (OECD, 1999). Emissions of carbon dioxide are the most critical factor for sustainability. This is emphasised in all major studies

(Steen *et al.*, 1999; Umweltbundesamt, 1997; OECD, 1999; Gorissen, 1996; Becker, 1998).

Sustainable Transport System: Principles and Definition

Today it is not clear, how a sustainable transport system will look like. Clear is that the setting of environmental protection targets can only be the beginning. In order to give a definition of environmentally sustainable transport it is necessary to understand the word sustainability in its somehow different meanings: instead of representing a state, a vision, or a future situation, it can also stand for a process or a path to be followed (Becker, 1998). As Cary writes: 'Sustainability is not a fixed ideal, but an evolutionary process of improving the management of systems, through improved understanding and knowledge. Analogous to Darwin's species evolution, the process is non-deterministic with the end point not known in advance' (Cary, 1998).

For this path nine sustainable transportation principles – the so called Vancouver principles – have been developed (Environment Canada, 1996). The first principle dealing with access and mobility describes the overall goal of transportation. Principles two to six delineate those social considerations that are necessary to meet when developing the transport system. Principles seven and eight correspond directly to carrying capacities and thus to ecological considerations. Principle nine describes how a sustainable transport system should be integrated in a market economy. The nine principles are:

- *Access* People are entitled to reasonable access to other people, places, goods and services.
- *Equity* To meet the basic transportation-related needs of all people, including women, the poor, the rural, the disabled, and children, nation states and the transportation community must strive to ensure social, interregional, and intergenerational equity. Developed economies must work in partnership with developing economies in fostering practices of sustainable transportation.
- *Individual and Community Responsibility* All individuals and communities have a responsibility to act as stewards of the natural environment, undertaking to make sustainable choices with regard to personal movement and consumption.

- *Health and Safety* Transportation systems should be designed and operated in a way that protects the health (physical, mental, and social well-being) and safety of all people, and enhances the quality of life in communities.
- *Education and Public Participation* People and communities need to be fully engaged in the decision-making process about sustainable transportation and empowered to participate.
- *Integrated Planning* Transportation decision makers have a responsibility to pursue mere integrated approaches to planning. They must involve partners from relevant sectors such as environmental, health, energy, financial, and urban design.
- *Land and Resource Use* Transportation systems must make efficient use of land and other natural resources while preserving vital habitats and maintaining bio-diversity.
- *Pollution Prevention* Transportation needs must be met without generating emissions that threaten public health, global climate, biological diversity, or the integrity of essential ecological processes.
- *Economic Well Being* Taxation and economic policies should work for, and not against, sustainable transport. Market mechanisms must account for the full social, economic, and environmental costs, both present and future, in order to ensure users pay an equitable share of costs.

Based on the above nine principles and with a focus on environmental or ecological sustainability the OECD Environment Directorate (1999) defines an environmentally sustainable transport system as one where

> transportation does not endanger public health or ecosystems and meets needs for access consistent with (a) use of renewable resources below their rates of regeneration, and (b) use of non-renewable resources below the rates of development of renewable substitutes (OECD, 1999).

The expression 'does not endanger ecosystems' includes that the integrity of the ecosystems is not significantly threatened, and that potentially adverse global phenomena such as climate change are not aggravated.

Criteria for Sustainable Transport Systems

Based on the description of today's unsustainable transport system and the mostly negative trends for future transport as well as on the definition of an environmentally sustainable transport system, sustainability criteria can be derived. Transport specific sustainable criteria for climate change, ozone (and photochemical smog), carcinogens, oxides of nitrogen and volatile organic compounds, noise, waste and disposal, nature and landscape protection as well as for urban living conditions (including the improvement of residential surroundings) are listed in Table 5.

The most important goal is the reduction of CO_2 emissions. The German Enquete-Commission on 'Protecting the Earth's Atmosphere' referred to earlier recommended an 80 per cent reduction of CO_2 emissions in OECD countries from 1990 to 2050. The German Federal Government decided to reduce CO_2 emissions in Germany by a total of 25 per cent between 1990 and 2005. At the end of 1999, the new German 'red-green' coalition government confirmed this goal of reduction, but did not declare which part of this reduction would have to be fulfilled by the transport sector.

The easiest way to distribute the burden of reduction by sector is to allocate the same target to all sectors. This would mean that the transport sector, and accordingly each transport mode, would need to reach the same reduction target, i.e. that of 25 per cent.

Towards a More Sustainable Transport System

If the reduction target for climate change and other sustainability criteria are fixed, the question arises as to what can be done to reach these targets. First, the external costs (as far as they are tangible) should be internalised. This is of major importance to meet the polluter-pays principle. The internalisation of external costs is, of course, no guarantee that a sustainable transport system is reached, nevertheless there is consensus that making transport services deliberately more expensive would help to achieve the specified targets with respect to environmental quality.

Beyond this, Walter and Spillmann (1999) provide a list of measures and options which could pave the way to a more environmentally sustainable transport system:

- *Technical Optimisation* Means of transport and fuel should be technically optimised.
- *Increase of Efficiency* Transport infrastructure should be used more efficiently, the means of transport should be working to full capacity.
- *Modal Shift* The share of the environmental efficient means of transport should be increased.
- *Avoidance* Access to goods, services and social activities should be reached with a lower amount of motorised traffic.
- *Slow Down* To reduce energy consumption, all means of transport have to be slowed down. This would involve the introduction and enforcement of road traffic speed limitations and the setting of upper limits for high speed trains. On the other hand, the speed of regular rail and ship freight transport would have to increase to guarantee their competitiveness.
- *Co-ordination/Integration* A sustainable transport system should be supported through the co-ordination of all stakeholders.
- *International Co-operation* Working together on an international level and especially within the European Union should be increased.
- *Information* The public ought to be provided with information and education about the consequences of today's traffic and about the use of a sustainable transport system.
- *Research/Development* Support of research as well as of pilot projects from administration, companies and private organisations for testing sustainable transport.

Table 8.5 Proposal for Transport Specific Sustainability Criteria

Protected Good	Environment Goal	From	to	Source of the criteria
Climate change:	–25% of the CO_2-emissions	1990	2005	Federal Government
Carcinogens (health)	–90% of the benzene, polycyclic aromatic hydrocarbons (PAH) and diesel particulates	1988	2005	SRU 1994 LAI 1992
	–99% of the benzene, PAH and diesel particulates		2010	SRU 1994 LAI 1992
Tropospheric ozone, secondary particles (health, crops)	–80% of the NO_x- and VOC-emissions	1987	2005	SRU 1994
	–40% of the NO_x- and VOC-emissions (Ozone-alarm)	Now		BimSchG §40
Noise protection (health)	Reduction of noise level to ≤ 65 dB (A) days ('protection of health')	Now	2005	UBA
	Reduction of noise level to ≤ 59 dB (A) t. / ≤ 49 dB (A) n. (in residential areas)		2010	UBA SRU 1994
	Reduction of noise level to ≤ 50 dB (A) t. / ≤ 40 dB (A) (in residential areas)		2030	AKWir
Conservation of Resources	Durability of automobiles and their parts		1997	BMU 1994
	Material before energetic utilisation rate of utilisation at least 85% of the total weight 95% of the total weight		2000 2005	BMU 1994 BMU 1994
Nature and Landscape	Implementation of the UMK Action plan 'protection of nature and traffic' e.g. conservation of areas with low volume of traffic	Now		UMK 1992
	No further extension or new roads, railways or waterways in valuable areas	Now		UBA
	Review of the extension of the road net (Bundes-wegeplan).	1998		UBA
	Applying the principle 'build new – build back' in infrastructure planning.	Now		UBA
Improvement of Residential surroundings (examples)	Reduction of life time accident risk for pedestrian/cyclist to be killed (1: 2.500) or to be injured to (1: 125)		2005	UBA
	Street including parking lane not more than 50% of the total road. (sidewalk width: 3,80m, cycle-path width: 2,50m		2005	Approach of the city of Berlin 1993

Source: Based on Gorissen, 1996

The Role of Cost-Benefit Analysis

The guidelines and proposals advanced by the Second Assessment Report of the Intergovernmental Panel on Climate Change (IPCC, 1995) about the role of cost-benefit analysis in evaluation apply to transport as well. Cost-benefit analysis provides an analytical framework that seeks to compare the consequences of alternative policy actions on a quantitative rather than a qualitative basis. The internal and most of external costs and benefits can be analysed by using the traditional cost benefit analysis which requires that all costs and benefits are expressed in a common monetary unit.

Concerning the climate change issue or the valuation of impacts on nature and landscape (bio-diversity), costs and benefits of transport have been estimated in several studies but the results are highly uncertain and controversial. Thus, other techniques such as cost-effectiveness analysis or multi-criteria analysis may be used for the inclusion of new transport-specific sustainability criteria. These techniques are included in modern cost benefit analysis (see Figure 8.3).

Figure 8.3   Modern Cost-Benefit Analysis (CBA)

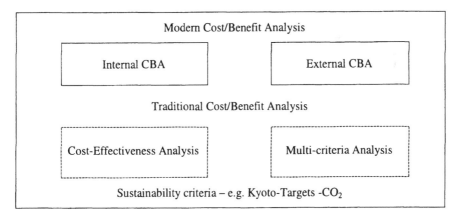

The purpose of this approach is to integrate the climate change issue which is the main impediment on the way towards a sustainable transport system into economic assessment. This should be done in the most cost-effective way.

Cost-Effectiveness Analysis   As cost-benefit analysis began to be applied to much broader fields, and particularly to the comparison of alternative portfolios of projects and to road policy choices, the increasing complexity made it necessary to keep the level of benefits constant and to analyse the problem simply in terms of finding the most effective or 'least-cost' option to meet the desired level of benefits. This has the additional advantage that benefits need not always be explicitly valued. For example, in power sector planning, models are applied to identify an capacity expansion plan minimising the present value of system costs fulfilling exogenously specified policy targets. This is the variant of cost-benefit analysis that has seen the most widespread application in the climate change discourse. There as with electricity power plants, the objective is to identify the least-cost option to achieve given levels or targets of greenhouse gas emission reductions, without any explicit attempt to specify what the benefits of that level of emission reduction may be (Munasinghe *et al.*, 1996).

Shadow Values/Avoidance Cost Estimates Another evaluation method is the use of shadow values. The concept of shadow values is also known as avoidance cost or mitigation cost approach. The methodology is similar to cost-effectiveness analysis. Environmental standards are derived from exogenously specified policy targets. Then policy options are identified which meet these environmental standards in the most cost-efficient way. An example is the calculation of mitigation costs of greenhouse gases due to national or international obligations (reduction of x per cent until year y).

The use of shadow values cannot be recommended as a first-best-solution from the perspective of welfare economics. It represents a kind of cost-efficiency analysis which is included in modern cost-benefit analysis. However, the approach is theoretically correct under the assumption that the selected reduction targets represent people's preferences appropriately. Shadow values estimate the opportunity costs of environmentally harmful activities assuming that a specified reduction target is socially desired. Shadow values can therefore be recommended as a second-best-solution in cases where damage costs are not available or not reliable.

A follow up of the well known study of IWW/Infras (1995) on behalf of the UIC has presented avoidance cost and damage cost results for climate change. The given ranges for Germany are avoidance costs between 35 to 37 Euro/t (IWW and Infras, 1999). The underlying 5.2 per cent reduction target represents the EU average agreed at the Kyoto process. For a reduction of 25 per cent the range is estimated at between 30 to 70 Euro/t CO_2.

Multi-Criteria Analysis The most basic requirement for the application of cost benefit analysis is that both costs and benefits are tangible. This is typically a two-step process: first, the costs and benefits are quantified in terms of the physical measures that apply; second, those physical impacts are valued in economic terms. Some applications of valuation techniques are likely to be controversial. Putting a value on human health and illness has been a major problem in the practical application of cost-benefit analysis in the past, even in a situation where one can agree on the levels of increased morbidity and mortality that might be caused by some policy or project. Efforts to place economic value on the loss of bio-diversity are extremely difficult. Recognising this problem has led to the development of so-called multi-criteria analysis techniques. These are expressly designed to deal with multiple objectives, of which economic efficiency may be only one. According to Munasinghe *et al.* (1996), multi-criteria analysis is a particularly powerful tool for quantifying and displaying the trade-offs that must be made between conflicting objectives. The advantage of multi-criteria analysis is that it does not need a common measuring unit, i.e. money. The disadvantage is that valuation in several dimensions, which are made comparable by scoring systems, are often not transparent.

Conclusions

With reference to the questions raised in the introduction of this chapter, we can draw the following conclusions, for sustainability criteria in general and for the transport system in particular:

- Sustainability criteria can be formulated in physical or monetary terms. Both physical and monetary criteria require normative judgements which have to be made transparent.
- In practical policy making, performance indicators like PSD and DSR have been preferred by international institutions like the OECD and the UN. These concepts avoid to use normative assumptions. But they should not be confused with sustainability criteria for which additional evaluation methods are needed.
- Operational sustainability indicators, as far as they are already available, mainly refer to environmental issues. These indicators can be used for practical policy making. However, due to the fact that

environmental aspects have to be weighted against economic and social goals, other concepts have to be used additionally.

- Cost-benefit analysis can offer important information about trade-offs between economic and ecological goals. But traditional cost-benefit analysis has been mainly developed for the valuation of local and regional projects with limited time horizons. As far as sustainability problems with their global and intragenerational problems are concerned, conventional cost-benefit-analysis is of limited value. As far as possible, aspects of intra- and intergenerational fairness must be considered in economic valuation studies. Modern cost-benefit-analysis must be understood in a broad to include cost-effectiveness analysis and multi-criteria analysis.

Annex 1 Driving Force – State – Response Indicators

Category	Chapter of Agenda 21	Driving Force Indicators	State Indicators	Response Indicators
Social	3. Combating poverty	Employment rate (in %) Ratio of average female wage to average male wage (in %)	Population living in absolute poverty (number and in %) Gini coefficient of income	
	5. Demographic dynamics and sustainability	Population growth rate (in %) Net migration rate (people/year)	Population density (people per square kilometre (km^2))	Total fertility rate
	36. Promoting education, public awareness and training (including gender issues)	Rate of growth of school age population Primary school enrolment ratio (in %) Secondary school enrolment ratio (in %)	Adult literacy rate (in %) Population reaching grade five of primary education (in %) Mean number of years of schooling	Gross domestic product (GDP) spent on education (in %) Female per 100 males in secondary school (number) Women per 100 men in the labour force (in %)
	7. Promoting sustainable human settlement development (including traffic and transport)	Rate of growth of urban population (in %) Transport fuel consumption per capita (litres) Number of megacities (population 10 million or more)	Population in urban areas (in %) Area and population of marginal settlements (km^2, number) Cost/number of injuries and fatalities related to natural disasters (US dollars, number) Floor area per person (square metres, m^2) House price-to-income ratio	Expenditure on low cost housing (US dollars) Expenditure on public transportation (US dollars) Infrastructure expenditures per capita (US dollars) Housing credit portfolio
Economic	2. International cooperation	Real GDP per capita growth rate (in %) Exports of goods and services (US dollars) Imports of goods and services (US dollars)	GDP per capita (US dollars) Environmentally adjusted domestic product (EDP) per capita/environmentally adjusted value added US dollars	Investment share in GDP (in %) Participation in regional trade agreements (yes/no)

	4. Changing consumption patterns	Depletion of mineral resources (in % of proved reserves) Annual energy consumption per capita (dollars)	Share of manufacturing value added in GDP (in %) Export concentration ratio (in %)	Proved mineral reserves (tons[t]) Proved energy reserves (oil equivalents) Life-time of proved energy reserves (years) Share of natural-resource intensive industries in manufacturing value added (in %) Share of manufacturers in merchandise exports (in %)	Ratio of consumption of renewable resources to that of non-renewable resources (in %)
	33. Financial resources and mechanisms	Ration of netresource transfer/GDP (in %)	Total official development assistance (ODA) given or received (in % of GDP) Dept/GDP (in %) Dept/service/export (in %)	Environmental protection expenditure as in % of GDP Environmental taxes and subsidies as in % of government revenue Amount of new or additional funding for sustainable development given/received since 1992 (US dollars) Programme of integrated environmental and economic accounting (yes/no) debt relief
	34. Transfer of environmentally sound technology, cooperation and capacity building			

Environmental				
Water	18. Protection of the quality and supply of freshwater resources	Annual withdrawals of groundwater and surface water as in % of available water Domestic consumption of water per capita (cubic metres (m³))	Groundwater reserves (m³) Concentration of faecal coliform in freshwater bodies (number/100 millilitres (ml)) Biochemical oxygen demand (BOD) and chemical oxygen demand (COD) in water bodies (milligrams (mg)/l)	Waste-water treatment (in % of population served, total and by type of treatment)
	17. Protection of the oceans, all kinds of seas and coastal areas	Catches of marine species (t) Population growth in coastal areas (in %) Discharges of oil into coastal waters (t) Releases of nitrogen and phosphorous into coastal waters (t)	Deviation in stock of marine species from maximum sustained yield level (MSY) (in %) Ration of MSY abundance to actual average abundance (in %) Algae index	Participation in maritime treaties/agreements (yes/no)
Land	10. Integrated approach to the planning and management of land resources	Land use	Area affected by soil erosion (km²)/erosion index	Land reform policy (yes/no) Decentralised local-level natural resource management (yes/no)
	12. Managing fragile ecosystems: combating desertification and drought	Fuelwood consumption per capita (m³) Livestock levels per km² in dryland Population living below poverty line in dryland areas	Area affected by desertification (km²)/desertification index Drought frequency	
	13. Managing fragile ecosystems: sustainable mountain development			
	14. Promoting sustainable agriculture and rural	Use of agricultural pesticides (t/km²)	Area affected by salinization and waterlogging (km²)	Cost of extension services provided and cost of agricultural

	development	Use of fertilisers (t/km²) Arable land (hectares (ha) per capita) Irrigation of arable land		research (US dollars) Area of land reclaimed
Other natural resources	11. Combating deforestation	Deforestation rate (km²/annum) Annual roundwood protection (m³)	Timber stocks (m³) Forest area (km²) Wood consumption as in % of energy consumption	Reforestation rate (km²/annum) Protected forest area as in % of total land area
	15. Conservation of biological diversity		Threatened, extinct species (number)	Protected area as in % of total land area
	16. Environmentally sound management of biotechnology			
Atmosphere	9. Protection of the atmosphere	Emissions of carbon dioxide (CO_2)(t) Emissions of oxides of sulphur (SO_x) and oxides of nitrogen (NO_x)(t) Consumption of ozone-depleting substances (t)	Ambient concentrations of sulphur dioxide (SO_2), carbon monoxide (CO), oxides of nitrogen (NO_x), ozone (O_3) and total suspended particulates (TSP) in urban areas (parts per million)	Expenditure on air pollution abatement (US dollars) Reduction in the emission of CO_2, SO_x and NO_x (in % per year)
Waste	21. Environmentally sound management of solid wastes and sewage-related issues	Generation of industrial and municipal waste (t)	Waste disposed/capita (t)	Expenditure on waste collection and treatment (US dollars) Waste recycling rates (in %) Municipal waste disposal (t/cap.) Waste reduction rates per unit of GDP (t/year)
	19. Environmentally sound management of toxic chemicals			
	20 and 22. Environmentally sound management of hazardous and radioactive waste	Generation of hazardous waste (t) Imports and exports of hazardous waste (t)	Area of land contaminated by hazardous waste (km²)	Expenditure on hazardous waste treatment (US dollars)

	wastes		
Institutional	35. Science for sustainable development		
	37. Cooperation for capacity-building		
	8, 38, 39 and 40. Decision-making structures	Mandated environmental impact assessment (EIA) (yes/no) Programmes for national environmental statistics and indicators for sustainable development (yes/no) Sustainable development strategies (yes/no) National councils for sustainable development (yes/no) Main telephone lines per 100 inhabitants (number)	Ratification of international agreements related to sustainable development (number) Number of local government employees per 1,000 of population (number) Personnel expenditure ratio (proportion of recurrent expenditure) spent on wage costs (in %)
	Strengthening of traditional information (part of 40)	Representatives of indigenous people in national councils for sustainability (yes/no) Existence of database for traditional knowledge information (yes/no)	
	Section III (23-32). Strengthening the role of major groups	Representatives of major groups in national councils for sustainability (yes/no)	

Source: United Nations, Economic and Social Council (1995)

Notes

1   This is not to say that monetary indicators are incompatible with the paradigm of strong sustainability. Prerequisite is that they are not separated but supplemented by physical indicators expressing threshold values of critical ecology.

2   As Gregor (1995) writes, a mass balance can be compared with the picture of a balance, where protons are put onto one pan, and the net sum of the system-related rates of protons consuming and producing processes, on the other pan. The critical value is the load holding both pans in an equilibrium.

3   Similar definitions can be found in Pearce and Atkinson (1993) and Pearce, Hamilton and Atkinson (1996). It should be added that critical elements of natural capital can neither be substituted by man-made capital nor by other elements of natural capital, as the example of the ozone layer and the panda bear may illustrate.

4   Most of the critical arguments pointing out the limits of traditional cost-benefit-analysis can be found in the IPCC Second Assessment Report (IPCC, 1995, WG III).

5   The structure of the IPCC includes three Working Groups: Working Group I (WGI) assessed the science of climate change. Working Group II (WGII) focused on the analysis of impacts and response strategies. Working Group III (WG III) studied the socio-economic implications of impacts, adaptation, and mitigation and prepared future emissions scenarios (Arris, 1996). Each Working Group prepared a final report and a summary for policy-makers (SPM) (IPCC 1995, WG I-III). The summaries are supplemented by a synthesis report covering the issues of all the three Working Groups (IPCC, 1995).

6   Additionally, some efforts have been made towards a more dynamic modelling of climate change damages which will not be discussed within this article. See for details Tol (1996a, 1996c).

7   Relative high (market) discount rates of six or more per cent normally represent the concept of opportunity costs of capital.

References

Arris, L. (1996), 'The IPCC Second Assessment Report: A Review', in *Environment Watch Western Europe* (EWWE), Special Edition March 1996, Cutter Information Corp.

Arrow, K.J., Parikh, J., Pillet, G. *et al.* (1996), 'Decision-Making Frameworks for Adressing Climate Change', in Bruce, J.J., Lee, H. and Haites, E.F. (eds), *Climate Change 1995 – Economic and Social Dimensions of Climate Change*, Cambridge University Press, Cambridge, pp. 53-78.

Arrow, K.J., Cline, W.R., Mäler, K.G. *et al.* (1996), 'Intertemporal Equity, Discounting and Economic Efficiency', in Bruce, J.J., Lee, H. and Haites, E.F. (eds), *Climate Change 1995 – Economic and Social Dimensions of Climate Change*, Cambridge University Press, Cambridge, pp. 125-144.

Azar, C. and Sterner, T. (1996), 'Discounting and Distributional Considerations in the Context of Global Warming', *Ecological Economics*, 19, pp.169-184.

Bayer, S. and Cansier, D. (1996), 'Methodisch abgesicherte intergenerationelle Diskontierung am Beispiel des Klimaschutzes', *Diskussionspapier der Eberhard-Karls-Universität Tübingen*, Wirtschaftswissenschaftliche Fakultät, Tübingen.

Becker, U. (1998), 'Principles of Sustainable Mobility and Guidelines for Nowadays Decision', Paper presented at the Eighth World Conference on Transport Research, July 12-17, Antwerp.

Bernow, S., Biewald, B., Dougherty, W. and White, D. (1996), 'Counting the Costs: Scientific Uncertainty and Valuation Perspective in EXMOD', in Hohmeyer, O., Ottinger, R. and Rennings, K. (eds), *Social Costs and Sustainability*, Springer, Berlin, pp.200-231.

Cary, J. (1998), 'Institutional Innovation in Natural Resource Management in Australia: The Thriumph of Creativity over Adversity', in *Abstracts of the Conference Knowledge Generation and Transfer: Implications for Agriculture in the 21st Century*, University of California-Berkeley, June 18-19, 1998, pp.11-13.

Cazeba Gutés, M. (1996), 'The Concept of Weak Sustainability', *Ecological Economics*, 17, pp.147-156.

CEC/US Joint Study on Fuel Cycle Costs (1993), 'Assessment of the External Costs of the Coal Fuel Cycle', Draft Position Paper prepared for DG XII of the Commission of the European Community, Brussels.

CEPN, ETSU, Ecole des Mines, IER, Metroeconomica (1994), 'Externalities of Fuel Cycles – ExternE Project', Summary report, European Commission, DG XII, Brussels.

Daly, H. (1990), 'Towards Some Operational Principles of Sustainable Development', *Ecological Economics*, 2, pp.1-6.

Daly, H.E. (1992), 'Allocation, Distribution and Scale: Towards an Economics That is Efficient, Just and Sustainable', *Ecological Economics*, 6, pp.185-193.

De Boer, B. and Bosch, P. (1995), 'The Greenhouse Effect: An Example of the Prevention Cost Approach', Paper prepared for the Second Meeting of the London Group on National Accounts and the Environment, Washington, March 15-17, 1995, Statistics Netherlands, Environment Statistics, Voorburg.

Environment Canada (1996), Proceedings of the OECD International Conference Towards Sustainable Transport in Vancouver, Canada, Hull.

Eurostat (1997), *Indicators of Sustainable Development*, European Communities, Luxembourg.

Fankhauser, S. (1995), *Valuing Climate Change – The Economics of the Greenhouse*, Earthscan, London.

Fankhauser, S., Richard, S. and Tol, J. (1995), 'The Social Costs of Climate Change: The IPCC Second Assessment Report and Beyond', Instituut voor Milieuvraagstukken (IVM), Frije Universiteit Amsterdam, Discussion Paper W-95/34, Amsterdam.

Fankhauser, S., Richard, S., Tol, J. and Pearce, D.W. (1996), 'The Aggregation of Climate Change Damages: A Welfare Theoretic Approach', Unpublished Paper, Free University Amsterdam, Institute for Environmental Studies, Amsterdam.

German Bundestag (ed.) (1991), 'Protecting the Earth – A Status Report with Recommendations for a New Energy Policy', Volume 1, Economica Verlag, Bonn.

290 *Project and Policy Evaluation in Transport*

Gorissen, N. (1996), 'Konzept für eine nachhaltige Mobilität in Deutschland', Schriftenreihe der Deutschen Verkehrswirtschaftlichen Gesellschaft e.V. Viertes Karlsruher Seminar zu Verkehr und Umwelt – Wege zu einer ökologisch verträglichen Entwicklung des Verkehrs, Karlsruhe.

Gregor, H.D. (1995), 'Das Critical Loads/Level-Konzept, ein ökosystemarer Ansatz für Umweltindikatoren auf der Basis von Wirkungsschwellen', *Umweltgeologie Heute*, Vol. 5, pp.51–58.

Hueting, R. and Bosch, P. (1991), 'Note on the Correction of National Income for Environmental Losses', in Kuik, Onno, Harmen and Verbruggen (eds), *In Search of Indicators of Sustainable Development*, Dordrecht, Boston, pp.29-38.

IPCC – Intergovernmental Panel on Climate Change (1995a), 'Summary for Policy Makers: Economic and Social Dimensions of Climate Change', IPCC Working Group III, IPCC Secretariat, Geneva.

IPCC (1995b), 'Second Assessment Synthesis of Scientific-Technical Information Relevant to Interpreting Article 2 of the UN Framework Convention on Climate Change', IPCC Secretariat, Geneva.

IPCC Working Group I (1995c), 'Summary for Policymakers of the Contribution of Working Group I to the IPCC Second Assessment Report', IPCC Secretariat, Geneva.

IPCC Working Group II (1995d), 'Summary for Policymakers: Impacts, Adaptation and Mitigation Options', IPCC Secretariat, Geneva.

IPCC Working Group III (1995e), 'Summary for Policymakers: Economic and Social Dimensions of Climate Change', IPCC Secretariat, Geneva.

IWW/Infras (1995), *Externe Effekte des Verkehrs* (Banfi, S., Gehrung, P., Gühnemann, A., Iten, R., Mauch, S., Rothengatter, W., Sieber, N. and Zuber, J), Study on Behalf of UIC, Paris.

IWW/Infras (1999), 'External Costs of Transport', Draft Working Paper and interim report on Methodology, Update for the IWW/Infras 1995 study on behalf of the UIC, Paris.

Krause, F., Bach, W. and Koomey, J. (1990), 'Energy Policy in the Greenhouse', Vol. 1 in *Warming Fate to Warming Limit: Benchmarks for a Global Climate Convention*, London.

Lee, R. (1996), 'Externalities Studies: Why are the Numbers Different?', in Hohmeyer, O., Ottinger, R. and Rennings, K. (eds), *Social Costs and Sustainability*, Springer, Berlin, pp.13-28.

Markandya, A. and Pearce, D.W. (1991), 'Development, the Environment, and the Social Rate of Discount', The World Bank Research Observer, Vol. 6, No. 2, pp.137-152.

Munasinghe, M., Meier, P., Hoel, M., Hong, S. W. and Aaheim, A. (1996), 'Applicability of Techniques of Cost-Benefit Analysis to Climate Change', in Bruce, J.J., Lee, H. and Haites, E.F. (eds), *Climate Change 1995 – Economic and Social Dimensions of Climate Change*, Cambridge University Press, Cambridge, pp.145-177.

Nagel, H.D., Smiatek, G. and Werner, B. (1994), 'Das Konzept der kritischen Eintragsraten als Möglichkeit zur Bestimmung von Umweltbelastungs- und qualitätskriterien – Critical Loads & Critical Levels', Nr. 20 der Materialien zur Umweltforschung des Rates von Sachverständigen für Umweltfragen, Stuttgart.

Nordhaus, W.D. (1991), 'To Slow or not to Slow? The Economics of the Greenhouse Effect', *The Economic Journal*, Vol. 101, pp. 920-937.

Norgaard, R.B., and Howarth, R.B. (1991), 'Sustainability and Discounting the Future', in Constanza, R. (ed.), *Ecological Economics: The Science and Management of Sustainability*, New York, pp. 87-101.

OECD Environment Directorate (1999), *Environmentally Sustainable Transport*, OECD, Paris.

Opschoor, J.B. and Reijnders, L. (1991), 'Towards Sustainable Development Indicators', in Kuik, Onno, Harmen and Verbruggen (eds), *In Search of Indicators of Sustainable Development, Dordrecht*, pp.7-27.

Pearce, D.W. (1993), *Economic Values and the Natural World*, Earthscan, London.

Pearce, D.W. and Atkinson, G.D. (1993), 'Capital Theory and the Measurement of Sustainable Development: An Indicator of Weak Sustainability', *Ecological Economics*, pp.103-108.

Pearce, D.W., Hamilton, K., Atkinson, G. (1996), 'Measuring Sustainable Development: Progress on Indicators', in *Environmental and Development Economics*, pp.85-101.

Pearce, D.W., Cline, W.R., Achanta, A.N., Fankhauser, S., Pachauri, R.K., Tol, R.S.J. and Vellinga, P. (1996), 'The Social Costs of Climate Climate Change: Greenhouse Damage and the Benefits of Control', in Bruce, J.J., Lee, H. and Haites, E.F. (eds), *Climate Change 1995 – Economic and Social Dimensions of Climate Change*, Cambridge University Press, pp. 179-224.

Rabl, A. (1993), 'Discounting of Long Term Costs: What Would Future Generations Prefer Us to Do?', Discussion Paper, Ecole des Mines, Paris, now published together with Rabl (1994) in *Ecological Economics*, Vol. 17, pp.137-145.

Rabl, A. (1994), 'Discounting and Intergenerational Costs: Why 0 May be the Appropriate Effective Rate', Discussion Paper Vol. 17 Ecole des Mines, Paris. Now published with Rabl (1993) in *Ecological Economics*, Vol. 17, pp.137-145.

Rennings, K. (1994), *Indikatoren für eine dauerhaft-umweltgerechte Entwicklung*, Metzler-Poeschel, Stuttgart.

Rennings, K. (1999), 'Linking Weak and Strong Sustainability Indicators: The Case of Global Warming', in Hohmeyer, O. and Rennings, K. (eds), *Man-made Climate Change – Economic Aspects and Policy Option*, Physica, Verlag, Heidelberg, pp.83-110.

Schellnhuber, H.J. (1995), 'Die internationale Klimawirkungsforschung auf ihrem langem Marsch zur Integrierten Modellierung', in Hennicke, P. (ed.), *Klimaschutz- Die Bedeutung von Kosten-Nutzen-Analysen*, Birkhäuser, Berlin, pp.52-82.

Scherp, J. (1994), 'What Does an Economist Need to Know About the Environment?', Economic Papers of the European Commission, No. 107, Directorate-General for Economic and Financial Affairs, Brussels.

Schmidt-Bleek (1994), *Wieviel Umwelt braucht der Mensch? MIPS – Das Maß für ökologisches Wirtschaften*, Berlin.

SRU Rat von Sachverständigen für Umweltfragen (Council of Environmental Advisors) (1994), *Umweltgutachten 1994 – für eine dauerhaft-umweltgerechte Entwicklung*, Metzler-Poeschel, Stuttgart.

Steen, P., Akerman, J., Dreborg, K.H., Henriksson, G., Höjer, M., Hunhammar, S. and Rigner, J. (1999), 'A Sustainable Transport System for Sweden in 2040', in Meersman, H., Van de Voorde, E. and Winkelmans, W. (eds), *World Transport Research*, Amsterdam.

Tol, R.S.J. (1996a), 'The Damage Costs of Climate Change: Towards a Dynamic Representation', *Ecological Economics*, Vol. 19, pp.67-90.

Tol, R.S.J. (1996b), 'The Damage Costs of Climate Change: Towards an Assessment Model and a New Set of Estimates', Draft Paper, Instituut voor Milieuvraagstukken (IVM), Frije Universiteit Amsterdam.

Tol, R.S.J. (1996c), 'The Climate Framework for Uncertainty, Negotiation and Distribution (FUND)', Discussion Paper 96/02, Institute for Environmental Studies, Free University Amsterdam.

Umweltbundesamt (1997), *Nachhaltiges Deutschland: Wege zu einer dauerhaft-umweltgerechten Entwicklung*, Erich Schmidt Verlag, Berlin.

United Nations Economic and Social Council (1995), 'Commission on Sustainable Development', Third Session, April 11-28, 1995, Item 3(b) of the Provisional Agenda.

van der Loo, F.A. (1993), 'From Sustainability to Indicator: The Climate Change Case', in Weterings, R. and Opschoor, J.B. (1994), *Towards Environmental Performance Indicators Based on the Notion of Environmental Space*, Report to the Advisory Council for Research on Nature and Environment, Rijswijk.

Walter, F. and Spillmann, W. (1999), 'Zwischenhalt auf dem Weg zum nachhaltigen Verkehr', GAIA, Vol. 8, No. 2, pp.93-100.

Weterings, R.A., Opschoor, J.B. (1992), 'The Ecocapacity as a Challenge to Technological Development', Advisory Council for Research on Nature and Environment, Rijswijk.

Wissenschaftlicher Beirat der Bundesregierung Globale Umweltveränderungen (German Advisory Council on Global Change) (1995a), *Jahresgutachten 1995 – Welt im Wandel: Wege zur Lösung globaler Umweltprobleme*, Springer, Berlin.

Wissenschaftlicher Beirat der Bundesregierung, Globale Umweltveränderungen (German Advisory Council on Global Change) (1995b), 'Scenario for the Derivation of Global CO_2 Reduction Targets and Implementation Strategies', Secretariat of the WBGU, Bremerhaven.

World Commission on Environment and Development (1987), *Our Common Future*, Oxford.

Printed and bound by CPI Group (UK) Ltd, Croydon, CR0 4YY

27/10/2024

01779859-0001